Mark Pinsent

SCIENCE for You

Physical Processes

Series editor

Lawrie Ryan

Published in 2002 by:
Nelson Thornes Ltd
Delta Place
27 Bath Road
CHELTENHAM
GL53 7TH
United Kingdom

02   03   04   05   06   /   10   9   8   7   6   5   4   3   2   1

A catalogue record for this book is available from the British Library

ISBN 0 7487 6695 2

Illustrations and page make-up by Wearset Ltd, Boldon, Tyne and Wear
Cartoons by Bede Illustration and Harry Venning

Printed and bound in Italy by Canale

Introduction

Science for You (Physical Processes) has been designed to help students studying Double or Single Science at Foundation level.

The layout of the book is easy to follow, with the key areas of physics split into four sections: Energy, Electricity, Forces and Waves. Each section has a title page which gives an outline of the topic area and the chapter headings.

Each chapter contains a set of double page spreads with every new idea on a fresh spread. Care has been taken to present information in an interesting way. You will also find plenty of cartoons and a few *unusual* examples to help you enjoy your work. Each new scientific word is printed in bold type and important points including useful equations are in yellow boxes.

There are short questions in the text, as well as a few questions at the end of each spread. These help you to check that you understand the work as you go along. The questions at the end of the chapter are there to encourage you to look back through the chapter and apply your new ideas. At the end of each section, you will find a selection of past paper questions to help with revision. These are on the coloured pages throughout the book.

At the end of each chapter, you will see a useful summary of the key facts you need to know. You can test yourself by answering Question 1 that follows each summary.

As you read through the book, you will come across these signs:

This shows where there is a chance to use computers to help you find information.

This shows where experiments can be done to support your work.

(The instructions are on sheets in the Teacher Support CD ROM.)

There is an extra section at the end of the book. Here you can get help with your Coursework, Revising and doing your exams, and Key Skills.

Using this book should make physics easier for you to understand. I hope you find the mixture of important physics concepts with some humour and 'real life' examples helps to bring you success in Science.

Good luck!

Mark Pinsent

Contents

Introduction to the world of physics

You are already an expert in physics!
You have been doing physics investigations ever since you were born. Just learning to sit up took you months of careful measurement and control of forces as you tried to avoid falling over.

Your science work at primary school would have included physics activities such as measuring forces to find the friction of shoes and reflecting light off mirrors while doing mirror writing.

During key stage three you will have continued to learn more about physics. You must have built a selection of electrical circuits using bulbs and batteries. You would have found out about the renewable and non-renewable energy sources. You might also have run across the playing field to measure your speed.

In all the physics work you have done, you have simply been trying to explain why things happen the way they do.
Physics is all about trying to understand the world around us.

Physics affects every part of our lives.
Next time you watch TV, remember the work of John Logie Baird.
He invented the television which made TV soaps possible.
Without Alexander Graham Bell inventing the telephone, who knows how long it would have taken to make mobile phones.
Sir Isaac Newton wrote a set of laws that defined the way forces affect the motion of objects. Two hundred years leater, Albert Einstein showed that Newton's Laws would not apply at very high speeds.

Good physicists don't accept things are as they are.
They look a little deeper to try to explain exactly what is happening.

Question   Why is a T-shirt red?

Normal answer   A T-shirt is red because of the dye used to make it.
                  True, but . . .

Physics answer A red T-shirt is red because it is absorbing
                  all the light that hits it except for the red light
                  which it reflects to your eyes.

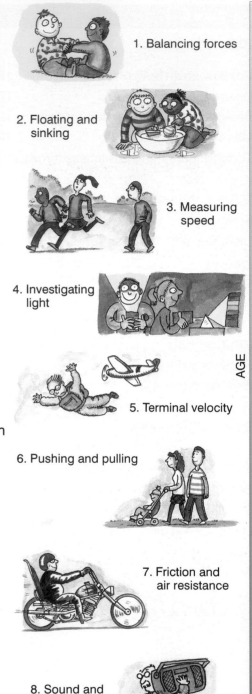

1. Balancing forces

2. Floating and sinking

3. Measuring speed

4. Investigating light

AGE

5. Terminal velocity

6. Pushing and pulling

7. Friction and air resistance

8. Sound and amplitude

Section One
Energy

In this section you will see how we transfer energy
from one type into another to make things work.
The world's energy sources will be described as well
as the processes we use to release energy from them.
You will be encouraged to think about the sources
of energy we will have to rely on in the future.
You will also learn how to calculate the amount of energy
transferred in a process.

CHAPTER 1

Energy transfer

▶▶▶ 1a Using calories/wasting joules

Have you ever 'run out' of energy after a day at school?
Why does the battery in a mobile phone have to be charged up?
Which has more energy, a family car travelling at 30 mph
or a piece of toast?

What is energy?

Energy is great stuff! We all know how it feels to have none left,
but what is energy in the first place? Is the type of energy in
a litre of petrol the same as the energy in a mobile phone battery?
Both batteries and petrol contain **chemical energy**. So when
your phone battery runs down, why would it be stupid to fill it with petrol?

Energy of toast = 400 kJ.
Energy of car = 150 kJ.

> **a)** Write a short description of what the word **energy** means to you.

Some facts to help

Energy is usually measured in joules (J).
Energy can be measured in other units:

    kilojoule (kJ) equal to one thousand joules, 1000 J
    megajoule (MJ) equal to a million joules, 1 000 000 J
    kilowatt hour (kWh) equal to 3.6 MJ or 3 600 000 J (household electricity bills use this unit (see page 52))

    ⎧ Calorie (cal) equal to 4.1 J   ⎫   (These two are older units of energy.
    ⎩ Kilocalorie (kcal) equal to 4100 J ⎭   We still use these units in foods.)

> **b)** People on a diet often 'count the calories'.
> Are they counting calories or kilocalories?

There are different types of energy: heat, light, sound, kinetic (movement),
chemical, nuclear, electrical and potential (stored).
They are explained in the next few pages of this chapter.

Our world only exists because energy changes happen.
Energy changes always follow the **Law of Conservation of Energy**,
as shown below.

> Energy can be **transferred/converted/changed** from one type to another.
> Energy can *never* be created (made) or destroyed (lost).

c) We often waste energy. However, the Conservation of Energy Law states that we can't destroy energy. It just gets spread out. Explain the difference between wasting and destroying energy using a few examples to help.

Energy **can be**:
- converted (changed) from one type to another;
- stored.

It is often wasted.

Energy **can't be**:
- destroyed;
- picked up;
- seen.

But you can see and feel the effects of it being changed.

> When energy changes occur, things happen.

It might be difficult to say exactly what energy is but energy changes play a vital part in our lives. Look at the two examples below:

- Chemical energy is stored within the food we eat. After digestion, sugars are released into our blood. They are taken to all parts of our bodies. These sugars are our fuel. We can think of them as 'burning up' within the cells of our body. This reaction is called respiration and it releases energy which our body uses.

- Electricity is generated in power stations (see Chapter 2). It is carried to homes through a network of pylons called the national grid. Then we use it to cook, give us light, for heating and to work all the machines in our homes.

Ears use energy to convert sound into signals for the brain.
Eyes use energy to convert light into signals for the brain.
Stomach uses energy to digest his snack.
Brain uses energy to convert signals from eyes and ears into sense.

Remind yourself!

1 Copy and complete:

Energy is measured in j......
The energy values of foods is often given in both kilojoules and ......
Energy is continually being ...... from one form to another.
Energy can never be ...... or ......
Energy is often ...... as it spreads out.

2 The food we eat is made up from the primary food groups. Protein is one group and it is used by the body for growth and tissue repair.
Find out which food groups provide us with most energy.

3 The energy on food packets is now given in both kJ and kCal.
100 g of a breakfast cereal contains 1600 kJ.
How many kCal is this?
(1 kCal is approximately 4 kJ.)

Energy changes affect every part of our lives.
It is impossible to understand energy transfers without
knowing about the different types of energy.

Heat energy

All materials are made out of tiny particles
(atoms and molecules). The temperature of
an object depends upon how much its particles are
vibrating in a solid, or moving around in the case of
liquids and gases.

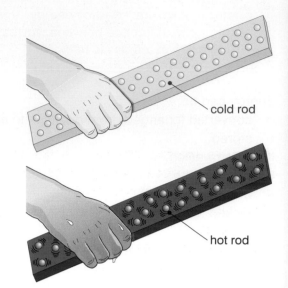
cold rod

hot rod

Heat energy always spreads from hot to cold.
As the object cools down, it will transfer its heat energy
(often called thermal energy) out to its surroundings by
either **conduction**, **convection** or **radiation** (see Chapter 3).

> **a)** If a solid is heated, it gains energy so that its temperature gets higher.
> The solid will expand.
> Describe what happens to the solid's particles.

Kinetic (movement) energy

Anything moving has kinetic energy.
The amount of kinetic energy an object has
depends on both its mass and its speed.

> **b)** A tennis ball was having a race with a 2 ton boulder
> down a hill, they were both travelling at 5 m/s
> Which one would want to stop? Why?

Sound energy

When objects vibrate, they cause particles
in the air to vibrate backwards and forwards.
The vibrating air particles form a sound wave.
When this wave reaches our ears, it also causes our
ear drum to vibrate. Our inner ear translates the
vibrating sound waves into messages for our brains.
We interpret these messages as sound.

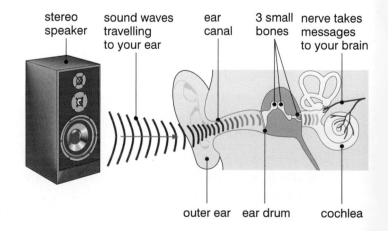
stereo speaker | sound waves travelling to your ear | ear canal | 3 small bones | nerve takes messages to your brain

outer ear | ear drum | cochlea

> The greater the vibrations in the air, the more sound
> energy we detect. The sound gets louder.

> **c)** Imagine you were listening to a mini disc player when your teacher
> was talking.
> How would they use sound energy to attract your attention?

Light energy Light rays entering our eyes get focused by our lens onto the retina. The retina converts the light into messages which our brain uses. Bright lights carry more energy. Light and heat both come from the Sun and together they are sometimes called **solar energy**.

lens   retina

message going up optic nerve

d) What process involving light energy and the colour green, is responsible for life on Earth?

Chemical energy When a firework is in its packet in the shop, what type of energy does it contain? We know it is made out of chemicals that will react but only when set alight. Think about a bar of chocolate. Eat it and you will get a boost of sugar to give you energy quickly.
So what energy does the chocolate have when it is in its wrapper? These are both examples of stored chemical energy.

I have the more stored chemical energy

But I am compact so I have better quality energy!

CRISPS

CHOCOLATE

e) When we light a firework, energy changes happen as the chemicals inside it react.
What are the different types of energy released once the firework is lit?

Remind yourself!

1 Copy and complete:

Hotter objects have more ...... energy than cooler ones.
Heat energy always flows from a ...... place to a ...... place as it spreads out.
Anything moving has ...... or ...... energy.
An object has kinetic energy because it is ......
The amount of kinetic energy also depends on the ...... of the object.

A road drill releases more ...... energy than a feather hitting the ground.
A floodlight gives more ...... energy than a bedside lamp.

2 Chemical energy is stored in foods and fuels. What would you call the energy stored in:

i)   a skier at the top of a ski run?

ii) a basket ball squashed on the ground just before it bounces back up?

▶▶▶ 1c More types of energy

Potential/stored energy

Foods, fuels and batteries are not the only things that can
store energy. We can store energy in springs,
stretched strings, flywheels and elastic bands.
Another form of stored energy is gravitational potential energy.

Gravitational potential energy (PE)

Imagine you are working under a book shelf.
A couple of books are just about to fall off and hit you on the head.
Which would you prefer to drop, a telephone directory
or a copy of *Science for You*?
Both books have **gravitational potential energy.**

> **Gravitational potential energy** is the energy stored within an
> object because of its distance above the ground.
> If given the chance, gravity will make the object fall.
> The potential energy is transferred into kinetic energy.

Which directory has more gravitational potential energy?

The telephone directory has more energy stored because it has
a greater mass. (But much more boring!)

a) The photo opposite shows people at the top of a sudden drop.
They are about to study the effects of extreme forces
after lunch and ice cream.
Their bodies have gained gravitational potential energy by
being lifted to the top of the ride.
What types of energy will their gravitational potential energy
be changed into as they fall?

Stretch an elastic band (catapult), or pull the firing pin on a
pinball machine. Both the elastic band and the spring have
energy stored in them (sometimes called **strain** or **elastic energy**).

If you let go of the catapult, the pellet flies through the air.
If you let the firing pin go on the pinball machine, the spring
inside it fires the pinball off and you score thousands of points
(hopefully).

b) Gravitational potential energy and the energy stored in a rubber
strap are both involved in bungee jumping.
Explain the energy changes that occur when a bungee
jumper jumps off a bridge.

Electrical energy

Electricity flowing through a complete circuit can cause things to happen.

Electrical energy is transferred to your CD player, making it work. Electric motors turn because of a flow of electricity. We use electric motors everywhere: microwave ovens, DVD players, personal stereos and electric tooth brushes. We also use electricity to cook food, heat houses, give us light (and most importantly, work mobile phones!).

radio
mobile phone
television
drill
food mixer
microwave
washing machine

c) Each of the devices in the picture box above changes electrical energy into other forms of useful energy.
Complete a table like the one below to include all of the devices shown above.

Name of device	Different types of energy produced when using device
Washing machine	Kinetic energy moving drum and water pumps, heat for water, sound
TV	

Nuclear energy

Nuclear energy is energy released from inside the atom.
Without nuclear energy, we would not be here!
We could not survive without the nuclear reactions that happen inside the Sun. The Sun provides us with nearly all of our energy.

We also use nuclear reactions on the Earth in nuclear power stations. They produce a small amount of the electricity that we use but most is produced by burning fossil fuels.

d) Most people do not believe that nuclear power stations are safe. They produce nuclear radioactive waste that can be very harmful to humans.
Does this mean that the electricity from a nuclear power station is more dangerous than other mains electricity? Explain your answer.

Nuclear power station.

Remind yourself!

1 Copy and complete:

Gravitational ...... energy (PE) is the energy stored in an object because of its ...... above the ground.
As the object falls, its ...... energy will be converted into ...... energy.
Electrical energy can easily be ...... into other forms of energy.

Electric lights change electrical energy into ...... energy and wasted ...... energy.
Nuclear reactions inside the ...... change nuclear energy into heat and kinetic energy.

2 Look at the sketches about stored/potential energy. Now draw your own set of cartoons to explain the different types of stored energy.

Changing and wasting energy

We have looked at the Law of Conservation of Energy and the different forms of energy. Now let's find out how to follow energy changes as they make things happen.

a) Write down the Law of Conservation of Energy in your own words. Try to do it without looking back to page 8.

So energy can never be destroyed, but we all know that it can be wasted. Think about a group of students talking in class instead of working. (They must be wasting energy!) Of course, if the teacher asks them, they always say that they are talking about the work. Look at the **energy transfer diagram** below for the students.

Chemical energy ⟶	Sound energy
from their food	all that talking

This is a very simple energy transfer diagram. We use energy transfer diagrams to trace the different types of energy involved when something happens in a process. We also try to identify any wasted energy that occurs as part of the process. In this case, the process is the pupils converting the **chemical energy** they gained from their lunch into **sound energy** as they talk.

Once the students have finished Science, they have a games lesson on the field learning to throw a javelin.

Here is their energy transfer diagram:

Chemical energy ⟶	Kinetic energy ⟶	Kinetic energy and wasted Sound energy	
from their food	Running and moving their arms to throw the javelin	The javelin's energy as it travels	They are all talking again as they wait

b) The second part of the students' games lesson was a 1500 m race. Draw an energy transfer diagram for the pupils highlighting any wasted energy.

Energy transfer diagrams

- Show how energy is transferred from one form to another.
- Show useful energy changes.
- Help us spot wasted energy changes.
- Are often called energy flow diagrams.

Sometimes we draw them in a single line:

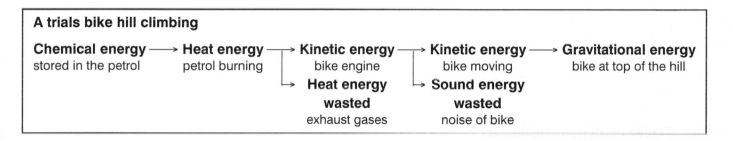

A mini disc player

Chemical energy ⟶ **Kinetic energy** ⟶ **Electrical energy** ⟶ **Sound energy** and wasted **Heat energy**
the batteries　　　　 the spinning disc　　 headphone wires　　　 the music　　　 disc player warms up

Sometimes they branch out:

A trials bike hill climbing

Chemical energy ⟶ **Heat energy** ⟶ **Kinetic energy** ⟶ **Kinetic energy** ⟶ **Gravitational energy**
stored in the petrol　 petrol burning　　 bike engine　　　 bike moving　　　 bike at top of the hill

　　　　　　　　　　　　　　　　　⟶ **Heat energy**　⟶ **Sound energy**
　　　　　　　　　　　　　　　　　　 wasted　　　　　 **wasted**
　　　　　　　　　　　　　　　　　 exhaust gases　　　 noise of bike

Sometimes they are drawn with arrows of different widths and with the values of energy given. This helps to show where the energy is being spread out.

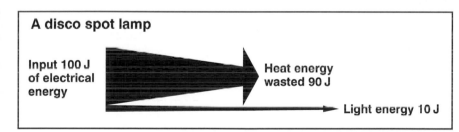

A disco spot lamp

Input 100 J of electrical energy　　　　Heat energy wasted 90 J

Light energy 10 J

Wasted energy can be any type of energy that we don't use, but is often wasted as heat energy.

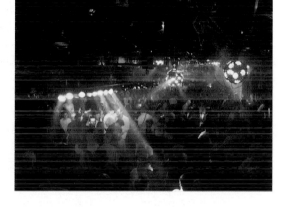

What's so good about electrical energy?

Think about five devices that you use at home.
How many of them rely on electrical energy to work?
The electrical energy we use comes either from the mains
supply or from batteries. The mains supply is used for
most household appliances. Batteries are only used
for portable devices such as personal stereos, torches
and mobile phones.
You can see how power stations generate mains
electricity in Chapter 2.

This will fix my rocket

a) Make a list of any electric devices that can use either the mains
or batteries for their energy supply.

Electrical energy is so widely used because it is very easy to
change it into other forms of energy.
- Electric fires produce heat;
- Electric lights make light;
- Stereo systems use electricity to make sound;
- Many household appliances produce kinetic energy e.g. CD player.

Example electrical energy transfer diagrams:

An electric fan heater

Electrical energy ⟶ **Heat energy** + **Kinetic energy** + wasted **Sound energy**
  from the mains   in the filament   in the fan   the fan can be noisy

A food mixer

Electrical energy ⟶ **Kinetic energy** + **Sound energy** + wasted **Heat energy**
  from the mains   the blades turning   hear it working   motor gets hot

A camera

Electrical energy ⟶ **Kinetic energy** + **Light energy** + wasted **Heat energy**
from stored chemical   the shutter, zoom lens,   from the flash   motor gets warm
energy in batteries   rewinding the film

---

b) Name a common form of transport often found in department stores.
It converts electrical energy into gravitational potential energy.

c) Look at the sketches below.
Create a table with these headings.

Device	Mains or battery	Energy transfer diagram

For each device, work out its energy transfer diagram.
Try to explain each step and include any wasted energy.

cordless drill

washing machine

television

laptop computer

hair tongs

hi-fi

hand-held games console

electric toothbrush

fan oven

door bell

microwave

Remind yourself!

1 Copy and complete:

Electrical energy is used by many household devices. It can be ...... into other ...... of ...... very easily.
Electrical devices often change electrical energy into ...... energy in stereos, ... energy in ovens, ... energy in spot lamps and ...... energy in machines that use a motor.

2 Electrical energy is often used in ski lifts.

Explain the energy changes on the ski lift.
Include an energy transfer diagram.

Summary

Everything that ever happens involves **energy changes**.

Energy changes can make things:

move/change temperature/give out sound/give out light

Energy is measured in **joules**, (J)

1000 J = 1 kilojoule (kJ)

Law of Conservation of Energy:

Energy can **never** be **created or destroyed**,
it can only be **converted** (changed) from one form to another.

The different types of energy:
- Heat (or thermal)
- Sound
- Chemical
- Electrical
- Kinetic
- Light
- Potential/Stored → Gravitational, → Strain or elastic
- Nuclear

As energy transfers happen, energy spreads **out** and becomes **less useful**.

Energy is normally **wasted** as it **spreads out**, often as **heat energy**,
warming the surroundings.

Energy changes can be traced using an energy transfer (or flow) diagram.

For example,
a portable games console:

Chemical energy →	**Electrical energy** →	**Light energy** + **Sound energy**
The batteries	Internal circuits	The display Mini speaker
		Heat energy
		Circuits and display get slightly warm

The direction of the arrows on an energy transfer diagram show how
the energy is being converted and spread out.

Electrical energy is used in many devices to give us **light, heat**, **sound**
or **kinetic energy**.

For example:

Light energy and heat energy from electric light bulbs.
Heat energy, kinetic energy and sound energy from an electric hair drier.

We can also use electrical energy to lift things,
giving them **gravitational potential energy**.

Questions

1 Copy and complete:

   i)   Energy can never be ...... or ......
        It can only be ...... from one type into
        another type.

   ii) The unit of energy is the ...... (J).
        ...... J is 1 kJ.

   iii) The different types of energy are ......, heat,
        ......, sound, ..., nuclear, ... and
        potential (or ......) energy.

   iv) When energy ...... out it becomes less
        useful.
        Energy is often ...... as heat energy.

   v)   Energy changes can be shown on an
        energy ...... diagram. The direction of the
        arrows shows how the energy is being
        t...... and spread out.

   vi) Electrical energy is easily ...... into other
        forms of energy such as ......, light, sound
        and ......
        When an electric motor ...... an object, it
        increases its gravitational ...... energy.

2 Use your ideas of energy changes
   and the Law of Conservation of Energy
   to explain the following.

   i)   The Sun never seems to change but it
        continues to give us energy.

   ii) The Earth is always receiving energy from
        the Sun in the form of light and heat but it
        never seems to get much warmer for long.

   iii) If you have a hot summer, your house gets
        nice and warm inside. In the winter you will
        still need to have the heating on.

   iv) A clockwork toy only walks about one
        metre before it stops.

   v)   There is no danger of a bungee jumper
        bouncing back up and hitting the platform
        they jumped from.

3 Imagine that at the dawn of time when the
   Universe had only just started there was
   100 000 000 000 000 000 000 000 J of energy.
   How much energy will there be left when the
   Universe finishes?

4 Look back at the different types of energy and
   their descriptions.
   Which type of energy do you think is the most
   important? Explain your answer.

5 Draw an energy transfer diagram for a plant
   growing.
   Start with the nuclear energy from the Sun.

6 Food chains are a type of energy transfer
   diagram.
   They show the direction that the energy passes
   from the original plant (the producer) to the end
   of the chain (the top carnivore).
   Work out a food chain that finishes with

   i)   A lion;

   ii) A student eating a bacon sandwich.

7 Draw energy transfer diagrams for the following
   energy changing processes.
   Highlight any wasted energy.

   i)   You riding a bike up a hill.

   ii) A down hill skier.

   iii) A solar powered calculator.

   iv) A formula one car as it brakes hard to go
        into the pits.

   v)   A child on a swing.

   vi) A toy doll that can walk and talk.

   vii) An electric kettle.

   viii) A coal fire.

   ix) A clockwork toy.

CHAPTER 2

ENERGY SOURCES

▶▶▶ 2a Our sources of energy

Where do we get all our energy from?

In Chapter 1 we looked at the different types of energy.
We saw how energy changes are everywhere in our lives.
In this chapter we are going to find out more about
the different sources of energy. We will look at how
they provide us with all the energy that we use.

> A **source** of energy is something that we can get energy from.
> A source of energy often releases more than one **type** of energy.

Natural gas contains chemical energy.
This is the type of energy that is stored in the gas.
If you burn gas it releases different types of energy,
mostly heat energy and light energy.

As it is easy to release energy from natural gas,
it is a very important source of energy.

a) Try to explain what these terms mean:
Types of energy
Sources of energy
Renewable energy sources
Non-renewable energy sources

Nearly all our energy sources can be traced back to the Sun:
- The Sun controls our weather systems affecting winds and rain;
- The Sun provides the energy to grow all our crops;
- Heat energy from the Sun can be used to provide electrical energy;
- Light energy from the Sun can also be used to provide electrical energy.

b) Explain how the Sun's energy would be passed to
you if you ate a burger with chips.

c) What is the link between the Sun and a piece of coal?

The Sun's role as energy provider

The formation of the fossil fuels **coal**,
natural gas and **oil** happened because
of the Sun's energy.
These fuels and **nuclear** fuels are:

> **Non-renewable energy sources:**
> only a fixed amount of energy is available,
> once we use them up, they have gone forever.

The Sun provides the energy for all plants to grow
They use the Sun's energy to make new material.
This process is called photosynthesis.
We call the plant material **biomass**.

The Sun controls our climate:
- Wind power gives us electricity.
- Rain over hills can be collected by dams to give us electricity.
- The Sun, as well as the Moon, has an effect on the size of our tides. We can use tidal power to give us electricity.
- The Sun's energy can also be used directly to give electricity.

Biomass and these sources of energy are known as:

> **Renewable energy sources:**
> energy sources that will continue to be available
> to use as long as the Sun lasts.

d) Find out and write down the word equation
for the process of photosynthesis.

e) The Sun has an effect on the size of our ocean tides
but what causes the tides?

Remind yourself!

1 Copy and complete:

Nearly all the …… energy comes from the Sun.
It is our main …… of energy.
Plants use the Sun's energy to grow during the
process of ……
The Sun is responsible for the Earth's weather
patterns including the …… and ……
…… energy sources such as solar power, ……,
and hydroelectricity transfer the Sun's energy
into ……

The non-renewable sources, the …… fuels, are
……, …… and natural …… They are also used
to give us electricity.

2 Where will we get all our energy from when coal,
oil and gas run out?

3 Represent all the energy transfers related to the
Sun on a spider diagram.

In Chapter 1 we saw how useful electricity is in all parts of our life, but where does it come from?

Electricity is often called a **secondary energy source**. That's because we have to use another source of energy to generate it. The electricity we use is generated in power stations. Most of these power stations use non-renewable sources of energy to produce electricity. The rest of our electricity is generated from renewable sources.

The numbers below give an estimate of the amount of each energy source currently being used in the world. (A large proportion is used to provide electricity.)

a) Construct a pie or bar chart showing this information:
Coal 25%, oil 40%, natural gas 25%, nuclear 7.5%, renewable sources 2.5%.

b) How do you think your chart would change if drawn in 50 years time?

Have you ever wondered how the petrol is lit in a petrol engine? The petrol is mixed with air and then ignited by a **spark**. The energy to produce the spark comes from the car's electrical system.

How do we generate electricity?

Cars use dynamos or alternators, which are types of generators. These all produce electricity. They are a bit like electric motors working backwards. (See Chapter 7 for more detail.)

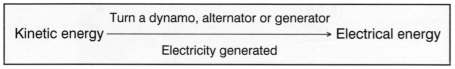

Kinetic energy $\xrightarrow[\text{Electricity generated}]{\text{Turn a dynamo, alternator or generator}}$ Electrical energy

c) A car has a battery to work the stereo and the mobile phone charger. (It also starts the engine.) So why does a car have an alternator?

White circle indicates alternator.

How do we turn a generator?

To generate electricity, we need to turn the dynamo, alternator or generator. Dynamos and alternators normally have pulley wheels and are turned by belts. Generators are turned by large 'propeller-like' fan blades called **turbines**.

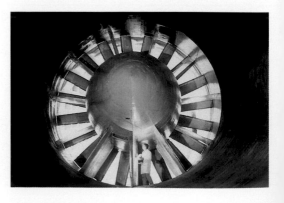

How do we turn a turbine?

Both non-renewable and renewable power stations use large turbines to produce mains electricity.

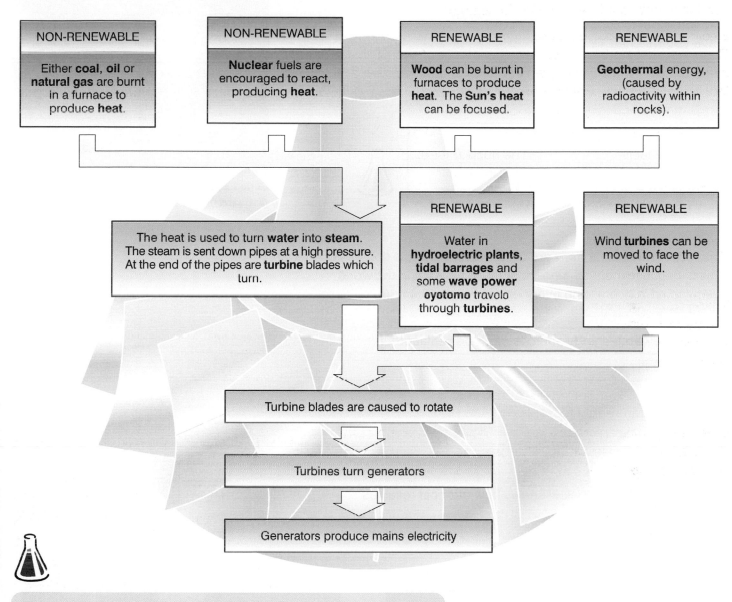

NON-RENEWABLE

Either **coal**, **oil** or **natural gas** are burnt in a furnace to produce **heat**.

NON-RENEWABLE

Nuclear fuels are encouraged to react, producing **heat**.

RENEWABLE

Wood can be burnt in furnaces to produce **heat**. The **Sun's heat** can be focused.

RENEWABLE

Geothermal energy, (caused by radioactivity within rocks).

The heat is used to turn **water** into **steam**. The steam is sent down pipes at a high pressure. At the end of the pipes are **turbine** blades which turn.

RENEWABLE

Water in **hydroelectric plants**, **tidal barrages** and some **wave power systems** travels through **turbines**.

RENEWABLE

Wind **turbines** can be moved to face the wind.

Turbine blades are caused to rotate

Turbines turn generators

Generators produce mains electricity

d) How does the electricity from power stations get to our homes?

Remind yourself!

1 Copy and complete:

...... is produced when a generator is ...... Generators are turned by ...... which are like large ......
Turbines in fuel burning power stations are driven by ...... released from boiled ......
Geothermal and ...... power stations also use steam driven turbines.

Flowing ...... or the wind is used to ...... turbines in many ...... stations.

2 Once a tree burns, it has gone!
Why are wood burning furnaces called renewable energy sources of electricity?

3 Using pictures to help, draw out your own version of the turbine flow diagram above.

23

When did you last use a non-renewable energy source?
Does your cooker or hob use gas?
Did you travel to school by car or bus?
Have you watched TV today?

We use non-renewable sources of energy for most transport.
We also use them to produce approximately 95% of our electricity.

*Camping stoves use butane
(a gas separated during fractional
distillation of crude oil).*

The non-renewable sources are:

The fossil fuels:
- Coal
- Oil
- Natural gas

The nuclear fuels:
- Uranium
- Plutonium

The fossil and nuclear fuels are called non-renewable sources of energy.
This is because the total amount of these fuels left in the world
is limited. It is decreasing every day as we use them up.

Non-renewable energy sources **can't be replaced**.

a) If you were President of the World, what would you do to make
the non-renewable resources last longer?

Coal

The picture flow chart below shows how our **coal** reserves
were formed.

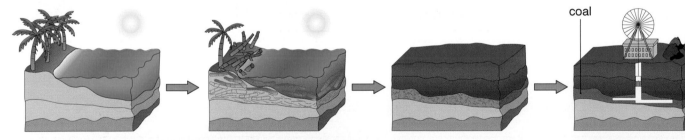

coal

300 million years later

We believe that the world has between 200 and 300 years
worth of coal left to mine. Much of it is hard to mine cheaply.
If we increase the amount we use, it won't last as long.

b) Trees are still dying and sinking into swamp regions of the Earth.
Why will this make very little difference to our energy crisis?

Oil and natural gas

These flow charts show how **crude oil** and **natural gas** were formed:

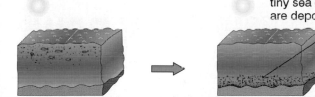 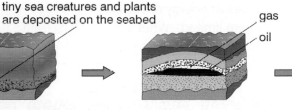 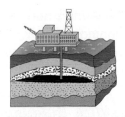

tiny sea creatures and plants are deposited on the seabed

gas

oil

150 million years later.

Coal is a common fuel used in some power stations.
We are now starting to build some smaller local power plants.
These use natural gas and refined oil instead of coal.
Natural gas is used for cooking and heating in many homes.
Refined crude oil has many other uses in our world.
These include petrol, diesel, aviation fuel and making plastics.

World reserves of natural gas are expected to last for
about another 35 years. Crude oil should last about 50 years.

> **c)** Dead sea animals which fed on sea plants are still sinking onto sea beds and getting trapped by sediments.
> What effect will this have on our oil and gas reserves?

Nuclear fuels

The fuel used in most nuclear reactors is the element **uranium**.
In some nuclear reactors, the element **plutonium** is released
as a 'by-product'. It is then used in other reactors as a fuel.
The supplies of uranium are expected to last between
60 and 100 years. Nuclear reactions are described in Chapter 16.

> **d)** Burning coal in power stations is noisy, messy and very smelly.
> Nuclear power stations are quiet and produce no air pollution (unless there's an accident!).
> Which type of power station would you like to live next to? Explain your answer.

Remind yourself!

1 Copy and complete:

......, oil and ...... are fossil fuels because they were formed from the remains of trees, plants and ...... millions of years ago.
The ...... fuels are the elements ...... and plutonium.
These fuels are known as ...... sources of energy, they are running out.
Once a non-renewable resource has been used, it has gone ......

2 i) List 10 things you could do in your life that would reduce the amount of electricity you use.

ii) If you manage to save energy yourself, would it affect the world energy crisis?

3 Find out how we get oil and natural gas from beneath the North Sea.

The world's non-renewable sources of energy are already running out now. Their supplies are limited. What will we do when they run out? We can't wait until fossil fuels run out completely before we find other energy sources.

We must act now!

We need to find *alternative* energy sources!

Renewable energy sources will not run out.
They are sometimes called **alternative** or **green** energy sources.
Nearly all renewable sources of energy depend on the Sun.

While the Sun lasts, we will always have energy.

a) What will happen to life on Earth when the Sun reaches the end of its life?

Energy from moving water

Many renewable energy sources work because of moving water.
Water is not an energy type or energy source.
Flowing water can **transfer** energy. It has kinetic energy (**KE**) which can be passed on to a water wheel or a turbine.

Hydroelectricity

We can collect rain water behind dams high up in mountains.
The water has gravitational potential energy (**PE**$_{grav}$).
When electricity is required in the National Grid, water is allowed to flow down the mountain. The water loses its PE$_{grav}$ as it falls, gaining more KE. At the bottom of the mountain the water flows through a turbine. Turbines turn generators and generators make electricity. See pages 22–23.

Tidal barrages

The Earth's tides depend mostly on the Moon's gravitational field. The Moon drags the oceans around behind it. As the tide 'comes in' the water level rises. Tidal barrages trap the high tide water behind a mini dam or large wall. When the tide goes out, the water level behind the barrage becomes higher than the new sea level. The trapped water is allowed to flow out to sea, through a turbine.

b) How would changes in tide height affect the electricity produced by a tidal barrage?

A giant water wheel transferring energy.

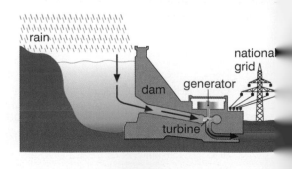

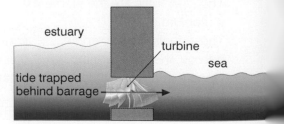

Wave energy

One type of wave generator uses moving water through a turbine to produce electricity. The wave is directed into a funnel near the shore. This turns it into a jet of water. The water jet is directed on to a turbine.

The other type of wave generator uses specially shaped floats. They sit on the water surface being pushed up and down by the passing waves. This constant movement is used to produce electricity.

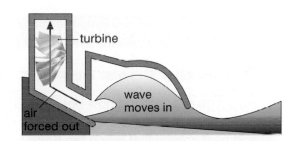

> c) How do you think we could transfer the electricity produced by a wave power plant out at sea to the shore?

Wind generators

Wind is caused by changes in air pressure in the Earth's atmosphere. Differences in pressure depend on the Sun's energy. We have used windmills for hundreds of years to grind wheat into flour for baking. Modern wind turbines generate electricity. They are often built in groups, forming a wind farm.

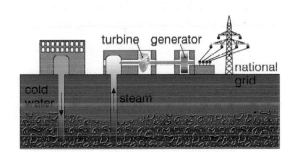

Using wave energy.

Geothermal energy sources

Geothermal energy does not depend on the Sun as it is heat energy released from inside the Earth. This energy comes from the **radioactive decay** of certain elements including uranium in the rocks (see page 204).

The geothermal energy is normally released as either hot water or steam. In hot springs the steam comes directly to the surface. In other systems we pump water down to the hot rocks and produce steam. Then we use the steam to turn turbines which drive generators.

Wind farm.

Geothermal energy.

Remind yourself!

1 Copy and complete:

...... sources of energy are sometimes referred to as the ...... energy resources.
Most of the ...... sources of energy get their energy from the ...... They will give us energy for as long as the ...... exists.
Geothermal energy is ...... released from radioactive elements.
......, tidal and ...... power all rely on moving water to turn ...... turbines drive ...... directly from the wind.

2 Electricity can be generated from burning domestic rubbish. Do you think this is a renewable or a non-renewable energy source?

3 Imagine that you live in a village in a valley. The government has decided to dam your local river for a hydroelectric plant.
Write a protest leaflet for or against the plans.

Using the Sun's energy

Have you ever noticed how hot it gets in a car that's been left out in the Sun? There are many ways of capturing the light energy and heat energy that reach the Earth from the Sun.

Solar cells capture the light energy from the Sun and convert it directly into electricity.
Have you got a calculator? Does it use solar cells?
We use them in many watches, calculators and other small devices. They are also used in space to power satellites.

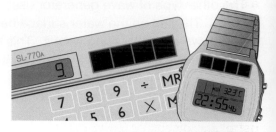

Solar panels can be fixed to the roofs of houses.
The Sun's energy warms water in the panels, which we then use inside the house for heating and washing. If you have ever had a shower in Greece, it is quite likely that the water was heated by either solar panels or inside a simple black tank on the roof.

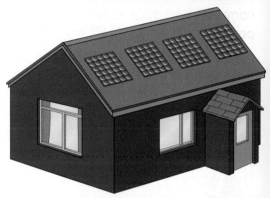

> **a)** Why are the water tanks on top of houses in Greece always painted black?

Solar energy reflectors use hundreds of mirrors to reflect the Sun's light and heat to a central focus. We use this energy to boil water, making steam. Scientists are working on the best way of using this system to generate electricity.

The Sun makes things grow: biomass

- When we eat cereals, vegetables, peanuts, bread or fruit we are getting the Sun's energy directly.
- When we drink milk, eat dairy products or eat meats, any fats in these foods can become stored energy for us. Animals ate the grass and we live off the animals.
- Solid waste from animals can be used to release **methane gas**.
- In Brazil **sugar cane** is used to make **alcohol**, used instead of petrol.
- **Wood burning furnaces** are being developed to replace burning fossil fuels. Burn a tree and replant with a new tree. Thirty years later you can do it all over again.

> **b)** Burning carbon-based fuels releases carbon dioxide. Why do we approve of burning wood but not fossil fuels, if they both release carbon dioxide?

The future energy sources and fuels

Hydrogen gas is being developed as a future fuel.
It is used in **fuel cells** that produce electricity.
The fuel cells require a steady supply of hydrogen
and oxygen to run. They work very quietly and produce
a safe waste product.

This car uses energy from fuel cells.

c) What waste produce do you think a hydrogen fuel
cell releases?

Fuel cells are being developed to power cars instead of
petrol and diesel. Many of the major car builders and oil
companies are now working together to make electric cars
powered by fuel cells a part of our future.

d) Why do you think oil companies would want to invest in
products that don't need petrol or diesel to run?

Giant solar cells in space could stay in orbit above the same
part of the Earth's surface (called a **geo-stationary orbit**).
They would be very large groups of solar cells able to collect
the Sun's energy without cloud problems. We would beam
down to Earth the energy they produce as **microwaves**.
We would convert the microwaves back into electricity.

Nuclear fusion is the type of reaction that happens inside the
Sun. It involves hydrogen atoms reacting together to produce the
element **helium**. We have not yet managed to build a fusion reactor.
If we ever do, all the world's energy problems will be solved!

> Hydrogen fusion to helium:
> Lots of energy released
> Plenty of hydrogen fuel available in water
> No pollution, and few radiation problems
> Let's get it sorted!

Remind yourself!

1 Copy and complete:

The ...... provides much of the world's energy.
Solar cells produce ...... from sunlight.
Solar panels ...... water for washing and heating.
Solar furnaces can be used to produce ...... for
driving ...... to generating electricity.
Hydrogen ...... cells produce ...... which can be
used to power a car.

2 Normal electric cars use lots of batteries. These
have to be charged up at night using electricity
from the mains.

Do you think that electric cars are a
green alternative to petrol cars?

3 Do some research into nuclear fusion reactors.
Try to find out why these reactors are so difficult
to build.

We have found out about the different sources of energy that we use.
How do they affect our environment?
Which source of energy gives us best value for money?
Which ones will you be using when you're 60 years old?

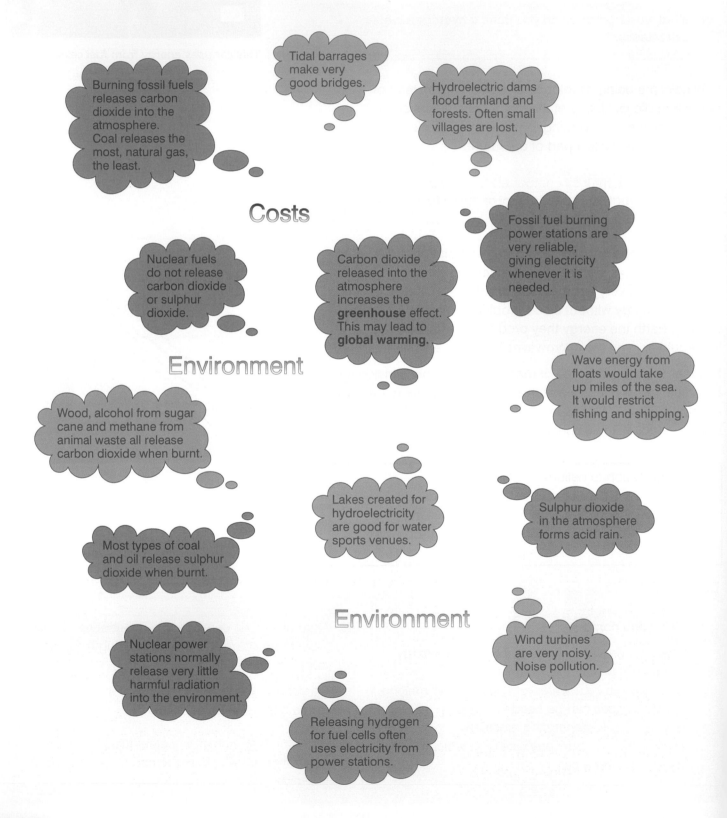

Tidal barrages make very good bridges.

Burning fossil fuels releases carbon dioxide into the atmosphere. Coal releases the most, natural gas, the least.

Hydroelectric dams flood farmland and forests. Often small villages are lost.

Costs

Nuclear fuels do not release carbon dioxide or sulphur dioxide.

Carbon dioxide released into the atmosphere increases the **greenhouse** effect. This may lead to **global warming.**

Fossil fuel burning power stations are very reliable, giving electricity whenever it is needed.

Environment

Wave energy from floats would take up miles of the sea. It would restrict fishing and shipping.

Wood, alcohol from sugar cane and methane from animal waste all release carbon dioxide when burnt.

Lakes created for hydroelectricity are good for water sports venues.

Sulphur dioxide in the atmosphere forms acid rain.

Most types of coal and oil release sulphur dioxide when burnt.

Environment

Nuclear power stations normally release very little harmful radiation into the environment.

Wind turbines are very noisy. Noise pollution.

Releasing hydrogen for fuel cells often uses electricity from power stations.

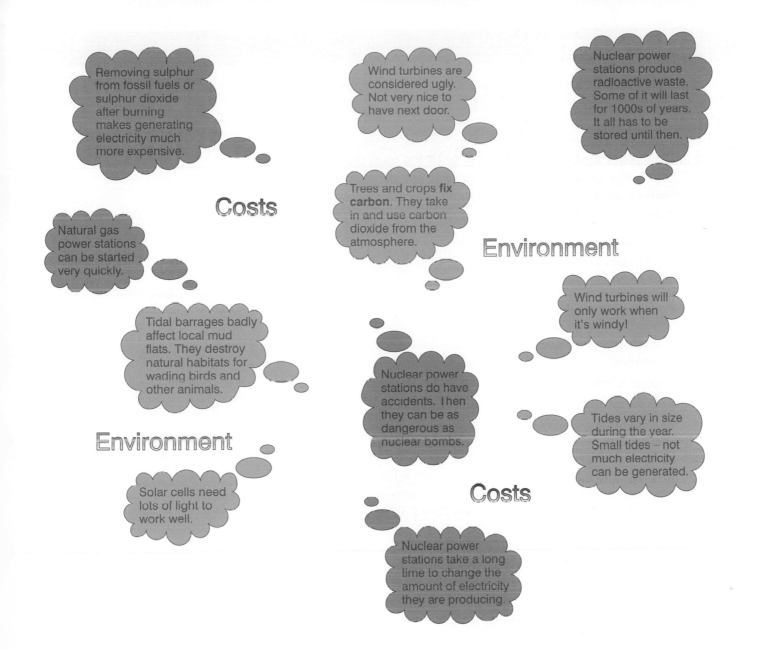

Removing sulphur from fossil fuels or sulphur dioxide after burning makes generating electricity much more expensive.

Costs

Wind turbines are considered ugly. Not very nice to have next door.

Nuclear power stations produce radioactive waste. Some of it will last for 1000s of years. It all has to be stored until then.

Trees and crops **fix carbon**. They take in and use carbon dioxide from the atmosphere.

Environment

Natural gas power stations can be started very quickly.

Wind turbines will only work when it's windy!

Tidal barrages badly affect local mud flats. They destroy natural habitats for wading birds and other animals.

Nuclear power stations do have accidents. Then they can be as dangerous as nuclear bombs.

Tides vary in size during the year. Small tides – not much electricity can be generated.

Environment

Solar cells need lots of light to work well.

Costs

Nuclear power stations take a long time to change the amount of electricity they are producing.

Processing Information

The information on these pages covers most of the renewable and non-renewable sources of energy.

Try either question 1 or 2 (do both if you have time)

1 You need to process this information. Working in pairs or alone, you should pick at least five different sources of energy. For each source, you should list in a table its good and bad points, those given here and any others you can think of. Pick the best source for our future world, balancing cost against environmental issues.

2 Working in a group of 4 or 6 students. Divide into two small groups. One group look at renewable energy sources, the other at the non-renewable sources. Use the information here to prepare and deliver a debate. Try to agree on the best and worst energy source for the future environment.

Summary

Nearly all the **Earth's energy** sources are related to the **Sun**.

The Sun's energy makes our crops grow:

The crops feed us.

They feed the animals that we then eat.

300 million years ago, the Sun's energy grew the trees that formed **coal**.

The Sun's energy grew the tiny sea plants that started the process **150 million years ago** that led to the formation of **oil** and **natural gas**.

The Sun affects our **weather systems**, resulting in **rain** clouds and **winds**.

The **non-renewable sources** of energy are:

the **fossil fuels**, coal, oil and natural gas

the **nuclear fuels**, uranium and plutonium.

Once used, these are gone forever!

The **renewable sources** of energy are:

- hydroelectricity
- tidal barrages
- wave generators
- wind generators
- solar energy
- biomass
- geothermal energy.

These sources of energy will last for as long as the Earth exists!

To generate electricity, we **turn a turbine** (like a propeller) which **turns a generator**.

We use fossil fuels, nuclear fuels and geothermal energy to produce steam.

The steam turns the turbines.

In hydroelectric power stations, flowing water turns the turbines directly.

In wind farms, the turbines are turned directly by the wind.

The energy debate is about the **cost** of each source of energy, how easy it is to use and its effect on our **environment**.

Non-renewables:

Provide most of our electricity quickly and with little fuss.

But: Burning fossil fuels releases CO_2 which leads to global warming.

Coal and oil can release **sulphur dioxide** when we burn them which causes **acid rain**.

Radioactive waste from nuclear power has to be stored for hundreds of years.

Renewables:

Are better for the environment (green).

But: They currently give us far less electricity.

Wind farms are noisy.

Hydroelectric schemes are hard to place and expensive to build.

Solar cells and panels don't work well when it is cloudy.

Tidal barrages change the habitats of many animals.

Questions

1 Copy and complete:

   i) The ...... is the Earth's energy provider.
   In the past, energy from the Sun was ......
   as ...... fuels were formed.
   Now the Sun's energy controls our ...... and
   grows our ......, providing us with all our
   food.

   ii) The non-...... energy resources are running
   out. Their use causes ...... leading to both
   global warming and ...... rain.

   iii) The ...... energy resources are less popular
   at present. They will continue to be
   available for as long as the ...... exists.

   iv) ...... are like big propellers.
   They are turned to drive ...... which produce
   electricity. ......, fast moving water or wind
   are all used to drive turbines.

2 The Sun's energy affects our climate.
Explain how the Sun:

   i) Causes rain over hills which then fills up
   dams used to supply hydroelectric power
   stations.

   ii) Causes pressure changes in the
   atmosphere resulting in winds which drive
   wind generators.

3 Write or draw a flow diagram to explain how
wood can be used in a wood furnace power
station to produce electricity.

4 Describe how coal was formed using the Sun's
energy. Use sketches to help.

5 Crude oil is used to generate electricity in some
power stations.

   i) List some more uses of crude oil.

   ii) When oil supplies run out, how will each of
   the uses you listed be served?

6 We need to find alternatives to burning fossil
fuels.

   a) Do some research into how we can get
   electricity from:

      i) Burning animal waste.

      ii) Burning domestic rubbish.

   b) Burning fuels releases carbon dioxide into
   the atmosphere. It is a gas that leads to
   global warming.
   So why is burning animal waste much better
   for the environment than burning coal?

7 A TV company have decided to create a new
'Castaway' experience.
You are to be part of the project but this time it
is set in a cut off valley in a range of ice-capped
mountains.
You need electricity to charge your mobile
phone batteries, play your music and work the
electric heating in the winter.
The TV company have given you the equipment
to build three different types of renewable
power plants.
Plan out your valley showing what renewable
power sources you will use and
where they would be placed.
Work out where the camp would
be and where the food would come from.
Use the valley sketch below to get you started.

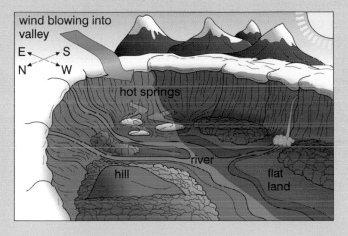

CHAPTER 3

THERMAL ENERGY TRANSFER

▶▶▶ 3a Heat energy and temperature

Heat energy is also called **thermal** energy.
Heat energy is spread out by conduction, convection or radiation. These are covered in the next few pages.

You probably know that all materials are made out of tiny particles.
The particles are called atoms and molecules.

In solids the particles do not move but can vibrate.
In liquids and gases the particles vibrate but they are able to move as well. They have kinetic energy.

Draw particle sketches of the following:

a) Warm solid close to melting   **b)** A cold liquid near to freezing

c) Liquid at boiling point   **d)** A hot gas

The amount of heat energy a substance has depends on its **internal energy**.

When heat energy is given to a solid;
the particles inside it vibrate more.
The solid has gained internal energy.

When heat energy is given to a liquid or a gas;
its particles vibrate more and gain kinetic energy by moving about faster.
The liquid or gas has gained internal energy.

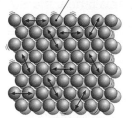

particles vibrate

solid

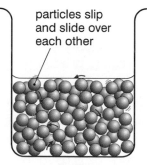

particles slip and slide over each other

liquid

The idea of hot and cold is something that you learnt when you were a toddler. People judge **temperature** by touch.
If something is hot, you move your hand away.
If too much heat energy flows through your skin, you get burnt!

Heat energy flows from hot to cold.

e) Picking up and holding which spoon would burn you the most?
A metal spoon in a pan on the hob warming food at 70°C or the same spoon left in the freezer at −20°C. (Skin temperature is approx 35°C.)

Temperature and internal energy

Are substances with large amounts of internal energy hot?
A sink full of water might have a temperature of 70°C.
An oven tray of chips would have a temperature of 200°C.

> **f)** Which do you think has more internal energy,
> the water or the oven tray?

Things at different temperatures exchange heat energy
and settle at a common temperature.
Put the oven tray into the water and it will cause a little steam.
After a few seconds the water and tray will settle at a new
temperature of about 80°C.
The water has increased its temperature by 10°C.
The oven tray has dropped its temperature by 120°C.
The water in the sink has more internal energy than the oven tray,
(even though it was at a lower temperature).

> **g)** Would the oven tray or the bowl burn your hand?

The oven tray would cause a bad burn. Think what would happen
if you picked it up without oven gloves. Heat energy would flow from 200°C
to your skin temperature at 35°C. You would 'cook your fingers'. Ouch!

> **Temperature** is a way of explaining how hot or cold something is.
> The **Celsius temperature scale** is used to compare different temperatures.
> Its units are °C (degrees Celsius).

A common older unit of temperature is the **Fahrenheit (F)**.
Scientists use the **absolute scale**. Its unit is the **Kelvin (K)**.

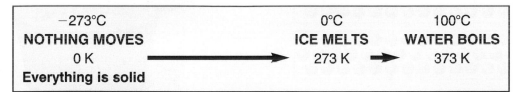

−273°C	0°C	100°C
NOTHING MOVES	**ICE MELTS**	**WATER BOILS**
0 K	273 K	373 K
Everything is solid		

Remind yourself!

1 Copy and complete:

Heat energy is also called ...... energy.
Everything is made out of tiny ...... which vibrate.
Increase the ...... of a substance and the
particles vibrate more. Particles in liquids and
...... are free to move. Increase their
temperature and they move ......
The amount of vibration and movement is the
...... energy of a substance.

2 Write your own definition of temperature.
Use the idea of internal energy and give some
examples of hot and cold objects to help your
explanation.

3 At absolute zero (0 K/−273°C) nothing
moves. You can't get any colder.

i) Find out about Lord Kelvin.

ii) Is there a highest possible temperature?

Have you ever:

- Walked on sand with bare feet on a hot summer day?
- Touched the hot tap immediately after running a bath?
- Tried to stir soup with a metal spoon that has been left in the soup while it was cooking?
- Wanted to pick up an ice-cold can and found that your fingers have stuck to it?

If so, you know what heat energy transfer by conduction is!

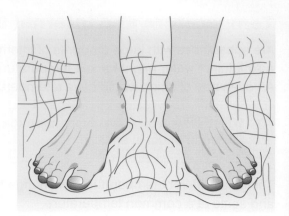

Sandy beach.

a) Explain the effects in each of the questions above. Talk about the flow of energy from hot to cold as part of your answer. Sketches would help.

When a solid is heated, its internal energy increases as its particles start to vibrate more.
The energy is passed through the solid from vibrating particle to vibrating particle.
Heat energy is **conducted** through the solid.

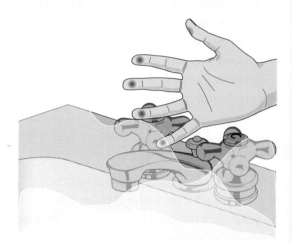

When a solid is heated, its particles vibrate more quickly. What do you think happens

b) to the size of the particles?

c) to the size of the solid?

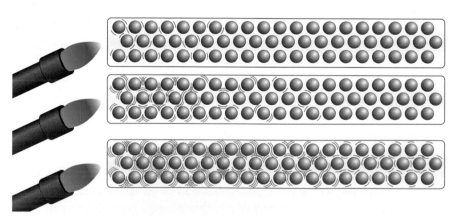

This diagram shows a glass rod as heat energy transfers through it. Gradually the particles are beginning to vibrate more. The heat energy is being transferred or **conducted** along the rod.

d) Conduction through the glass rod above is a slow process. Why do you think metals are much better and faster conductors of heat?

Conductors and insulators of heat

All solid materials fall somewhere on this line:

Insulators or very
poor conductors of heat
wood, polystyrene
plastics, ceramics

Very good
conductors of heat
lead, steel, copper, gold,
graphite and diamond

An **insulator** does not conduct heat.
Non-metal elements and compounds are mostly insulators.

Pure carbon (a non-metal) has two common structures,
diamond and graphite. Both conduct heat but only graphite
conducts electricity. Diamond conducts heat well because
it has a very rigid structure.
Graphite conducts heat because it has some free moving electrons.
Electrons are tiny negatively charged particles which orbit
the nucleus of atoms.

Diamond.

Graphite.

e) Explain why you can fry an egg on a rock in a hot country.

f) Explain why metal objects feel cold.

Free electrons

All metals conduct both heat and electricity. Lead is a worse
conductor of heat than aluminium. Copper is one of the best.
Metals conduct heat so well because they have a rigid structure
and also some free electrons. Free electrons are not attached
to any one atom. They are able to move about inside the metal.
As we heat up a metal the free electrons gain more kinetic energy
making them move faster. They knock into atoms and other
electrons spreading their energy out.

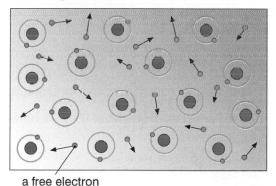
a free electron

Remind yourself!

1 Copy and complete:

Heat energy spreads through solids by ......
Solids are made out of ......, (atoms and
molecules). When heat energy is supplied, the
particles inside start to ...... faster, passing their
energy on to the ...... next to them.
The heat energy is ...... through the solid.
Metals have millions of ...... electrons.
The free electrons' ...... energy increases when
the metal is heated.

Energy is passed on to other ...... so heat
conducts through the ......

2 If glass and other ceramics are poor conductors
of heat, why are oven dishes often made out of
these materials?

3 Find out why diamond can conduct
heat so well but will not conduct
electricity.

ICT

Useful definitions

Heat energy travels in liquids and gases by convection.
To understand how heat travels by convection there are some
other important definitions we have to think about first.

Volume is the amount of space that a substance takes up.
It is measured in cubic metres (m^3).

Mass is the amount of particles of matter (or stuff) present
in the substance. It is measured in kilograms (kg) (see page 124).

Density is the amount of mass in one m^3 of the substance.
It is measured in kilograms per metre cubed (kg/m^3).

Lead is a very dense metal so a **small volume** of lead has a **high mass** for its size.
Polystyrene is a low-density solid so the same **small volume** has a very **low mass**.

Solids normally have quite a high density while liquids normally
have a lower density. Gases have an even lower density.

a) List some high- and some low-density solids.

b) Mercury would sink in water. It has a higher density than water.
List some substances that have a lower density than water.

Heat energy and density

Liquids or gases are called **fluids**. When they are heated,
they increase their volume (expand). The particles are not
held in position and are able to move about more quickly.

As the particles move about they hit other particles which
causes them to move about faster as well.
The particles now take up more space. So this part of the fluid
gets less dense.

c) Which part of a liquid do you think is the less dense,
the hotter or the cooler part?

cold   hot

Convectional currents

If you heat a pan of water, the hot water at the
bottom gets less dense. It will rise to the top.
The cooler water at the top moves down to take its place.

hotter particles rise
towards the surface

convectional
current

cooler particles
travel down

Movement of hot material to a colder place is called **convection**.
The motion that occurs is called a **convectional current**.

d) Copy and complete the table below.
Use your ideas about convectional currents to help.

Example of convection	Your explanation
Modern central heating radiators have welded fins pointing towards the wall.	
Glider pilots, already launched, are able to fly upwards without an engine.	
The top surface of the water in a bath is always hotter than the rest of the water.	
You often feel a draught when sitting near a door in a hot room.	
Hot water tanks fill up with cold water from the bottom.	

Remind yourself!

1 Copy and complete:

Thermal energy is carried through a ...... or gas
(often called ......) by......
As the fluid is heated, the ...... regions become
less ...... than the other parts of the fluid.
This causes them to ...... towards the ......
surface of the fluid. The hotter regions are
replaced by ...... more dense parts of the fluid
which ...... down to fill the gaps.
The constant flow of fluid, ... up and ......
down is called a ...... current.

2 Explain what is happening in the picture below.
Use the idea of convectional currents.

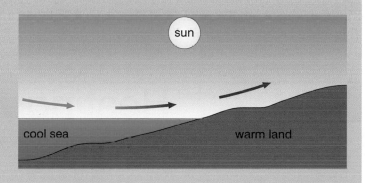

sun

cool sea

warm land

If you want to 'top up' the suntan on your right shoulder,
what do you do?

You sit up with your shoulder pointing towards the Sun.

How does a grill element cook food if hot air rises and the
food is always below the grill?

*The food is cooked by 'heat rays' travelling down from
the grill into the food.*

The heat rays are called **infrared rays**.

> **a)** If you stand near a bonfire, why does the part of your body
> facing the fire feel so much hotter than the part facing away?
>
> **b)** With barbecues, you must wait until the flames have died
> down before you put the food on.
> If you don't use the flames, how does the food cook?

Heat energy transfer by radiation ⟶ **waves** which carry heat energy.

The waves are called **infrared rays** or **infrared radiation**.
The hotter an object is, the more infrared rays it radiates.

Infrared waves are similar waves to **light waves**.
These are both **electromagnetic waves** (see Chapter 15).

Heat in – heat out

We have seen that atoms and molecules make up all materials.
If infrared radiation hits an atom, it increases its internal energy.
This makes it vibrate more.
If an atom starts to vibrate less, it gives out some infrared radiation.

Infrared radiation **absorbed** (is taken in)	**Heating** ⟶	Internal energy increases atoms/molecules **vibrate more**.
Infrared radiation **emitted** (given out)	⟵ **Cooling**	Internal energy decreases atoms/molecules **vibrate less**.

So how does the Sun give us heat energy?

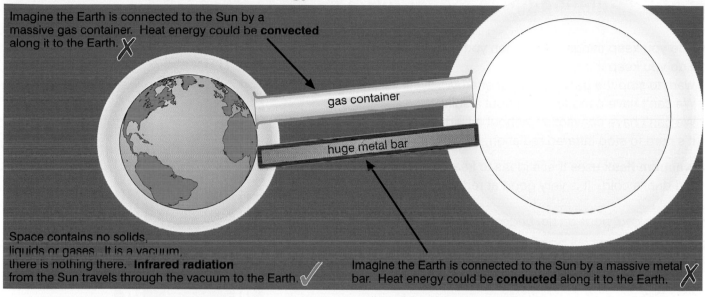

Imagine the Earth is connected to the Sun by a massive gas container. Heat energy could be **convected** along it to the Earth. X

gas container

huge metal bar

Space contains no solids, liquids or gases. It is a vacuum, there is nothing there. **Infrared radiation** from the Sun travels through the vacuum to the Earth. ✓

Imagine the Earth is connected to the Sun by a massive metal bar. Heat energy could be **conducted** along it to the Earth. X

c) Look at the three ideas above.
Explain why only radiation can transfer energy from the Sun to the Earth.

Properties of infrared radiation

- It is not a particle, it has no mass.
- It can travel through a vacuum.
- It behaves like visible light.
- It is reflected by shiny surfaces.
- It is absorbed by dull surfaces.

> **Good absorbers = Good radiators**
> Dark matt surfaces
>
> **Good reflectors = Poor absorbers**
> Light shiny surfaces

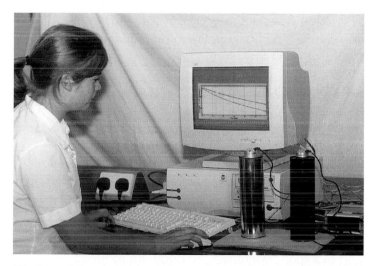

Black container radiates more heat energy than white container.

Remind yourself!

1 Copy and complete:

Heat energy transfer by waves is called ...... radiation.
When infrared radiation is ...... by an atom, it makes the atom vibrate more.
This increases the internal ...... of the material.
Infrared radiation can ...... through a vacuum and it behaves like visible ......
Dull dark surfaces ...... infrared radiation and shiny light surfaces ...... it.

The best ...... of infrared radiation are also the best emitters or radiators.

2 Explain what is happening in the experiment shown in the photo above.

3 Collect information for a mini project into how and where infrared radiation is used. Uses include TV remote controls and night vision cameras.

How do you keep things warm when you go on a picnic?
How do you keep things cold?
We want to stop the transfer of heat energy.

- We can't have conduction without a solid.
- We can't have convection without a fluid (liquid or gas).
- It's hard to stop infrared radiation, but we can reflect it back again.

The **vacuum flask** uses these ideas to keep hot drinks hot,
or cold drinks cold. It is very good at reducing heat losses.

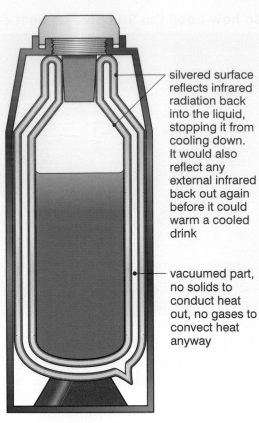

silvered surface reflects infrared radiation back into the liquid, stopping it from cooling down. It would also reflect any external infrared back out again before it could warm a cooled drink

vacuumed part, no solids to conduct heat out, no gases to convect heat anyway

A vacuum flask.

People use cool bags on hot beaches.

a) Explain how a cool bag can keep food and drinks cool.

b) Does a cool bag reduce conduction, and/or convection and/or radiation.

Insulating your body

Trapped air is a very poor conductor of heat.
When you wear clothes, it is not only the material that keeps you warm. You are kept warm by the air trapped within the fibres and between the layers.

c) When the weather is cold, is it better to wear a thick fleece or three thinner jumpers?
Explain your answer using ideas about heat energy transfer.

Modern anoraks and lightweight sleeping bags are made of fibre
insulation material. They have thin layers of reflective plastic strips
between the fibres.
By doing this, both conduction through the material
and radiation away from the body are reduced.

d) Draw a cross-section of some material cut from a lightweight sleeping bag. Label the insulation and reflective layers with clear notes explaining how each reduces heat energy loss.

Mountain rescue teams always include a silvered reflective
blanket in their survival kits. People in difficulties on mountains
get very cold. The thin blankets help to stop them getting any colder,
keeping them alive.

Some survival blankets have a plastic cover over the silvered layer.

Keeping warm at home

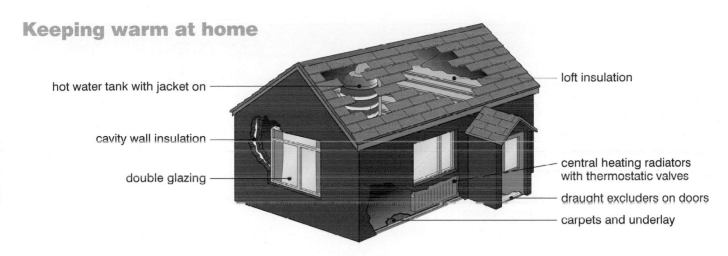

hot water tank with jacket on

cavity wall insulation

double glazing

loft insulation

central heating radiators with thermostatic valves

draught excluders on doors

carpets and underlay

Brand new homes have to have many of the features above included.
Building inspectors check all new homes to make certain
that they are properly insulated.
Older homes are often very poorly insulated.
The table below gives some rough estimates for the costs and paybacks
of improving a three bedroom house.

Method of heat loss reduction	Type of energy transfer reduced	Cost to install	Annual saving	Payback time
Loft insulation	Conduction/convection	£240	£60	4 years
Cavity wall insulation	Conduction/convection	£600	£80	7.5 years
Draught excluders	Convection	£40	£55	under a year
Double glazing	Conduction/convection	£3500	£50	70 years
Radiator controls	Convection	£120	£20	6 years
Hot water tank jacket	Conduction/convection	£12	£16	under a year

e) Which methods of thermal insulation have the quickest payback?

f) From which part of the house does the most heat energy escape?
Hint: think biggest savings, not installation cost.

g) Double glazing only saves £50 per year and is expensive to install.
Why do people still get their houses double glazed?

Remind yourself!

1 Copy and complete:

A ...... flask is designed to keep its contents at a steady ...... The vacuumed part stops ...... and ...... The silvered part stops ......
Clothes keep you warm by trapping ...... in the fibres and between layers.
Air is a very ...... conductor of heat.
Heat energy wastage within the home is reduced most by ...... and loft insulation.

2 Why do sea mammals have so much blubber under their skin?

3 Carry out a survey of the different methods of heat loss reduction used:

i) in your home,

ii) in a friend's home,

iii) in the oldest part of your school.

Summary

Heat energy is also called **thermal** energy.

Heat energy flows from hot to cold.

Temperature is a way of explaining how hot or cold something is.
The **Celsius temperature scale** is used to compare different temperatures.
Its units are degrees Celsius (°C).

Conduction is the process of heat energy transfer through solids.
Metals are very good conductors of heat.
Plastic, wood and glass are poor conductors of heat.
They are called **insulators**.
When a solid is heated, the particles inside it vibrate more.
The solid has gained internal energy.

Convection is the process of heat energy transfer in **fluids** (liquids and gases).
When heat energy is given to a fluid, its particles vibrate more.
The particles also gain kinetic energy, so they start to move about faster.
The fluid has gained internal energy.
The heated particles take up more space.
That part of the fluid becomes less dense and rises.
As it rises, other cooler parts sink.
The motion that occurs is called a **convectional current**.

Radiation is heat energy transfer by energy-carrying **waves**.
The waves are called **infrared rays** or **infrared radiation**.
The hotter an object is, the more infrared rays it gives out (radiates).
Dark matt surfaces are both good absorbers and good radiators
of infrared waves.
Light shiny surfaces are good reflectors and poor absorbers.

Insulation reduces heat energy transfers.
When you wear clothes, you are kept warm by the air trapped within
the fibres and between the layers. This reduces heat energy transfer
by conduction because trapped air is a very poor conductor of heat.
We can reduce heat losses by radiation by reflecting the infrared rays
back towards the heat source using silvered surfaces.

Reducing heat energy transfer is good for the environment.
It saves money and helps to conserve the world's energy reserves.
Home improvements which reduce heat energy losses include:
loft insulation, hot water tank jackets, cavity wall insulation, carpets and underlay,
double glazing, draught excluders on doors and radiator thermostatic valves.

Questions

1 Copy and complete:

i) Heat energy always flows from ...... to ......

ii) The ...... of an object is measured in degrees ( )

iii) Heat energy passes through solids by ......
As a solid is heated, its particles ...... more, passing energy onto nearby atoms.
...... are good conductors of heat while wood and ...... are poor conductors (I......).

iv) ...... is the process of heat energy transfer in fluids (...... and ......).
Particles that are ......, vibrate and move more.
The movement of particles, hotter ......, cooler ...... is called a ...... current.

v) Heat energy transfer also occurs by ......
Infrared ...... are given off by any object h...... than its surroundings. Dark ...... surfaces give off (emit) and take in (......) the most ...... radiation.
Shiny surfaces ...... infrared radiation.

vi) Insulating homes to r...... heat energy losses helps to conserve the ...... energy reserves.
Five methods of reducing ...... loss from houses are ......, ..., ..., ... and ......

2 Draw sketches for each of the circumstances described below. Show some of the particles inside the materials making it clear how they are vibrating and maybe moving as well:

i) A metal rod with one end held in a flame.

ii) A glass rod with one end held in a flame.

iii) A cup of tea as it is left to cool.

iv) A pan of boiling water on a hob.

3 i) Place the following materials into order from poorest to best conductor of heat?

Oak, copper, polystyrene, diamond, lead, silver.

ii) Would the order you worked out above be the same if the materials were being tested for electrical conductivity instead of heat?

4 Explain clearly with drawings where necessary, how convectional currents are responsible for each of the following:

i) The warmest part of a room is in the middle near the ceiling.

ii) A hot drink takes longer to cool down when it has a lid on it.

iii) A hot air balloon goes up higher when the burner is used.

iv) Washing dries best on a warm dry windy day.

5 Design a piece of equipment that could be used to warm up a drink using sunlight.

6 Explain each of the following using the principles of reflecting, absorbing and emitting infrared radiation:

i) Houses in hot countries are often white.

ii) Solar panels are always painted black.

iii) People normally wear white or very light colours in hot weather.

iv) Radiators are normally painted white.

7 You are going on a camping trip to the North Pole at Christmas. You want to spot Father Christmas as he sets off on Christmas Eve. Design a special suit to keep you warm while you spend the night out Santa watching. Label your design with explanations about the features that will keep your body insulated from the freezing weather.

CHAPTER 4

MEASURING ENERGY TRANSFERS

▶▶▶ 4a Work done – energy transferred

You are an expert at using energy. You use it every second of every day. We know that our food gives us energy. So we could work out how much energy we take into our bodies in a normal day.

How can we work out how much energy we use?

Equation 1 to learn

Transferring energy is often called 'doing work'.

work done = energy transferred

Every time you move, your muscles are **transferring energy**.

If you throw a ball, you have **done work** on it because you have given it **kinetic energy**.

When you sit down on a sofa, you transfer energy into its springs. You are doing work on the springs.

If you write, you are doing work. Not on the paper, on the pen!

a) List five things that you have done today since you left home. They will all involve transferring energy.

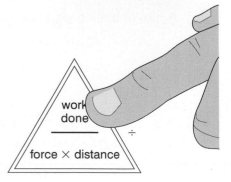

Equation 2 to learn

	work done	=	force applied	×	distance moved
	or energy transferred				in the direction of the force
Units	**joule (J)**		**newton (N)**		**metre (m)**

Equations or **formulas** can sometimes be put into a simple triangle to help you use them.
All you have to do is put your finger over the bit of information that you need to calculate. The other two parts left uncovered will tell you if you need to divide or multiply to get the answer.

Using the work done equation

Example 1 How much work do you really do?

Writing uses a force of about 0.5 N to press and drag the pen. The average word length is about 5 cm if all the letters are stretched out into a line. To write 200 cm (2 m) would be approximately 40 words. How much work would you have done?

work done = force applied × distance moved

= 0.5 N × 2 m

= 1 joule

> **1 joule** is the energy **transferred/work done** when a **force** of **1 newton** is applied (used) over a distance of **1 metre**.

b) Calculate how many words you would have to write to transfer 5 J.

c) If you were set a 1000 word essay for homework, how much energy would you use writing it?

Example 2 Is it easier to type the essay?

Writing a 1000 word essay would be the same as drawing a line about 50 m long. This would be about 25 J of work done.

When you type, the force on the keyboard is only about 0.2 N and the keys are pushed down about $\frac{1}{4}$ of a cm. On average there are five letters and a space per word. 1000 words is 6000 key strokes

0.25 cm × 6000 = 1500 cm or 15 m.

How much work would you have done?

Work done = Force applied × distance moved

= 0.2 N × 15 m

= 3 joules

d) How much energy do you use to type a 5000 word essay?

e) The energy in one cornflake is about 300 J.
How many words would the energy from one cornflake allow you to type?

Remind yourself!

1 Copy and complete:

Equation 1
...... done is the same as energy transferred.
When a ...... is applied over a ......, energy is transferred.

Equation 2
Work done = ... × ......
One joule of energy is transferred when a force of one ...... is applied over one ......

2 A student catches a bus to school.
The bus accelerates away from the bus stop with a force of 1000 N. It accelerates for the first 200 m. How much energy has been transferred?

3 A student cycling home has 10 000 J of kinetic energy. He has to stop in 20 m.

i) What force must his brakes apply?

ii) Why is the force he puts on the brake levers so much less than that needed to stop the bike?

Think about these questions:
- How many times do you climb a flight of stairs each day?
- Have you picked up a bag today?
- How much energy do you use when you lift a piece of food to your mouth?

a) What force are you working against when you lift a book on to a shelf?

When you 'do work' upwards you are increasing an object's **gravitational potential energy**.
When something falls, it is transferring its gravitational potential energy into kinetic energy. The amount of gravitational potential energy an object has depends on its **weight** and its **height** above the ground.

Weight is a force caused by the pull of **gravity**.
Gravity is a **mass attraction force**.
The Earth's mass is so great that it attracts all other masses on or near its surface towards its centre.
(This is explained in more detail on page 124.)

b) List five things that you have lifted today.

c) Look at your list for b). Arrange the different 'lifts' in order from least to most gravitational potential energy gained.

Equation 3 to learn

change in gravitational potential energy	=	**weight**	×	**change in vertical height**
Work done *upwards*, energy transferred *upwards*		a force		a distance moved *upwards*
Units joules (J)		newton (N)		metre (m)

d) Compare the gravitational potential energy equation with the work done equation on the previous page. What is the similarity between them?

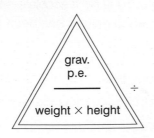

Using the gravitational potential energy equation

Example 1 Climbing stairs to classrooms

You could have climbed at least 10 flights of stairs today.
How much gravitational potential energy did you gain?

Average pupil's weight including a bag   600 N
Height of a set of stairs in school   3 m
Total height of climbing stairs 10 times   30 m

$$\text{change in gravitational potential energy} = \text{weight} \times \text{height}$$
$$= 600 \text{ N} \times 30 \text{ m}$$
$$= 18\,000 \text{ joules}$$
$$= 18 \text{ kJ}$$

(This is approximately the energy stored in a peanut).

e) A tray of empty beakers on a science lab trolley weighs 20 N.
    The tray was on the top of the 0.5 m high trolley before getting
    knocked off by a student.
    How much energy did the tray of beakers have just before
    it was sent crashing to the lab floor?

f) How much energy would you use lifting a 60 N bag 1.5 m
    on to your back?

Example 2 Lifting a workbag 24 times a day

How many times a day do you lift your bag and then drop it down again?
A school bag weighs 60 N.
The lift to your shoulder is about 1.5 m.
So lifting the bag 24 times in a day: total lift = 24 × 1.5 m = 36 m

$$\text{Change in gravitational potential energy} = \text{weight} \times \text{height}$$
$$= 60 \text{ N} \times 36 \text{ m}$$
$$= 2160 \text{ joules}$$
$$= 2.16 \text{ kJ}$$

g) How many cornflakes would you need to eat to carry out the work done in Example 2?
    (Remember, each contains about 300 J)

Remind yourself!

1 Copy and complete:

   When you ...... an object, you are doing ...... on
   it 'upwards'. This is increasing the object's ......
   potential energy.
   The increase in gravitational potential energy
   depends on an object's ...... and the ...... it is
   lifted by.

   Equation 3
   Change in gravitational ...... energy
   $$= \text{......} \times \text{...... in vertical height.}$$

2 A football weighs 2 N. It has 24 J of gravitational
   potential energy.
   How far above the ground is it?

3 A litre of pure water has a mass of 1 kg.
   Its weight on the Earth would be 10 N.
   Imagine a litre of water in a dam, 150 m above a
   turbine. How much kinetic energy would it have
   when it hit the turbine after being allowed to drop
   from the dam?

Power is a very useful word.
You probably use it all the time.
But what does it mean?

- You might want a more **powerful** stereo.
- Your friend's mobile phone might have a higher **power** rating than yours.
- You might prefer to use the most **powerful** drill in CDT.

a) Write down names of five things that you link with the words power or powerful.

> Power is the **rate** of doing work or the **rate** of transferring energy.
>
> It is how much energy is transferred **every second**.

A more powerful stereo transfers more sound energy every second.
A phone with a higher power rating will convert more energy from its battery every second.
The most powerful drill transfers more energy every second.
It is less likely to get jammed or overloaded.

Equation 4 to learn

$$\underset{\text{(watt, W)}}{\textbf{power}} = \frac{\overset{\text{(energy transferred)}}{\textbf{work done} \text{ (joules, J)}}}{\textbf{time taken} \text{ (seconds, s)}}$$

This equation is often written to calculate energy transfer instead:

energy transferred = power × time
(work done)

b) Look carefully at the 'triangle' equation on the right.
Write out the three possible equations it gives.
Check that two of them are the same as those in the yellow box above.

50

Using the power equation

Example 1 Pupils climbing more stairs to classrooms

It might take you 20 seconds to climb two flights of stairs
to a second floor classroom.
The gravitational energy that you would transfer could be 4000 J.
What would your power rating be?

$$\text{Power} = \frac{\text{energy transferred}}{\text{time taken}} \div$$

$$\text{Power} \div \frac{4000 \text{ J}}{20 \text{ s}} = 200 \text{ W}$$

You have a power rating of 200 W. You have transferred 200 J every second.

> The **watt** is the unit of power.
> 1 watt = 1 joule per second
> 1 W = 1 J/s

c) You climbed the same stairs as those above in double the time.
What would your power rating be?

d) If the staircase was 3 times higher, you would transfer 12 000 J
of gravitational potential energy climbing it.
What would your power be if you climbed it in 1 minute?

Example 2 Students need light to read

When you are reading your science notes late at night,
how much energy does the light bulb use?

Power rating of desk lamp = 60 W
Time spent reading = 3 minutes = 180 seconds

$$\begin{aligned}\text{Energy transferred} &= \text{power} \times \text{time} \\ &= 60 \text{ W} \times 180 \text{ s} \\ &= 10\,800 \text{ J} \\ &= 10.8 \text{ kJ}\end{aligned}$$

Remind yourself!

1 Copy and complete:

Power is the ...... of doing work. It is the amount
of energy ...... every second.

Equation 4
Power = ... ÷ time taken
This equation is often used in the form;
energy transferred = ... × ......
The ...... (W) is the unit of power.
One Watt is one ...... per ......
1000 ...... = 1 kW

2 A 150 W light bulb uses up 150 J of energy every
second.
How much energy does it use in:

i) 10 seconds

ii) 10 minutes

iii) 1 hour?

3 Do a survey of the power rating of a range of
household machines and other devices. They
normally have a stamp somewhere stating their
power rating.

Have you ever been told;
   'Stop leaving the lights on!'
    'Turn the TV off when you are not watching it!'
    'Stop wasting so much electricity!'?
How much electricity are you really wasting when you
leave a light on all night?

An electricity meter.

> **a)** How is the amount of electricity used in a house calculated?

Electrical energy transfer

Let's look at the amount of energy transferred by an electrical appliance.
We need to know its power rating and how long it is used for.
We use the power equation in the form below:

   Energy transferred = power × time   (Equation 4)

Example 1 2000 W electric fire used for 2 hours

Time fire used for = 2 hours = 120 minutes = 120 × 60 seconds
                                    = 7200 seconds
                   Power of fire = 2000 W
        energy transferred = power × time
                        = 2000 W × 7200 s
                        = 14 400 000 J
                        = 14.4 MJ

Just using one electric fire for only two hours uses 14 400 000 J.
Imagine how many zeros would be on a quarterly home electricity bill!
To make life easier for everybody, household electricity bills are
calculated using a much larger unit of energy, the **kilowatt hour** (**kWh**).

Equation 5 to learn

	electrical energy transferred	=	power	×	time
Units	kilowatt hour (kWh)		kilowatt (kW)		hours (h)

   1 kW = 1000 W (so just a larger unit of power)
    1 h = 3600 s (so just a larger unit of time)
1 kWh = 1 kW × 1 h

> **b)** How many joules are there in 1 kWh?
>   (Hint: put the watts and seconds values above into the equation.)
>
> **c)** Compare equation 5 with equation 4 on the last page.
>   What is the difference?

Using the electrical energy transferred equation

Example 1 (revisited) 2000 W electric fire used for 2 hours

The fire's power is 2000 W = 2 kW
The fire is used for = 2 h
electrical energy transferred = power × time
= 2 kW × 2 h
= 4 kWh (Less zeros than the 14 400 000 J!)

The fire used four domestic electrical units (4 kWh).
Electricity companies charge about 8p per unit (less for night rate users).

Equation 6 to learn

total cost = number of electrical units × cost per unit
(£ and pence) (number of kWh used) (pence each)

Example 2 Turn that light off!

What would it cost to leave a 100 W light bulb on for 10 hours?
time = 10 h
power = 100 W = 0.1 kW
electrical energy transferred = power × time
= 0.1 kW × 10
= 1 kWh (the number of units used)

Using equation 6
total cost = number of electrical units × cost per unit
= 1 kWh × 8p
= 8p to leave a 100 W light on all night!

d) How much would it cost to leave four 100 W light bulbs on for 10 hours?

e) Which would cost more to use, a 1 kW kettle for 15 minutes (0.25 h), or a 250 W (0.25 kW) television for an hour?

Remind yourself!

1 Copy and complete:

The amount of ...... energy used by an appliance depends on its ...... rating and how ...... it is used for.
Domestic electrical energy is calculated using a larger unit of ...... than the joule. It is called the ...... hour (kWh).

Equation 5
Electrical energy transferred (kWh)
= ...... (kW) × ...... (h)

Equation 6
Total ...... of electricity (£ and pence)
= number of ...... (kWh) × ...... of units (p)

2 Copy and complete this table:

Device	Power (kW)	Time (h)	Energy (kWh)	Unit cost (pence)	Total cost (pence)
Vacuum cleaner	1	2	2	8	16p
Light	0.1	15		8	
Oven	3	2		8	
Lawn mower		2		8	8p
TV	0.25	12		8	
Stereo	0.2			8	16p

▶▶▶ 4e Energy efficiency

Humans are always wasting energy.
You are probably quite good at wasting it yourself!

● Getting a lift to school; you want kinetic energy,
   but cars waste heat and sound energy.
● Cycling to school; you need kinetic energy,
   but you and the bike both waste heat energy.
● Mini CD players; you want sound energy,
   but the motor inside gets hot, wasting heat energy.

> **a)** Write out 2 or 3 energy transfer diagrams for things that you have done today. In each diagram, spot both the useful energy and wasted energy.
>
> **b)** For the activities you described in a):
>   i) Which wasted the most energy?
>   ii) How could you have reduced the amount of energy wasted?

In Chapter 1 you saw how energy wasted during energy transfers spreads into the surroundings.

> **c)** What form of energy is the most frequently wasted?
>
> **d)** What effect does this have on the temperature of the surroundings?

Efficiency

We use the idea of **efficiency** to compare energy transfers.
Processes or machines that are **efficient** waste less energy.
They convert more of their input energy into useful output energy.

Equation 7 to learn

$$\text{Efficiency} = \frac{\text{useful energy transferred by device}}{\text{total energy supplied to device}} \div$$

Gives efficiency as a decimal fraction (*always less than one*)

This equation is often used in one of these forms:

$$\text{Efficiency} = \frac{\text{useful output energy}}{\text{total input energy}} \times 100\%$$

$$\text{Efficiency} = \frac{\text{power output}}{\text{power input}} \times 100\%$$

These forms of the equation give efficiency values as percentages.

> You can never have a process that is more than 100% efficient.

Using the efficiency equation

Example 1 A student walking across the shopping centre

A student decides to walk 500 m back across a shopping centre
to save 50p on a new music CD.

Force on ground to walk = 45 N

Distance walked   = 500 m

Using   work done = force applied × distance moved

                    = 45 N × 500 m

          energy used = 22 500 J *This would be **useful** or **output** energy.*

The student used up 112 500 J of energy from his burger
and chips lunch to walk to the music shop. This would be her **input** energy.

$$\text{Efficiency} = \frac{\text{useful energy transferred by device}}{\text{total energy supplied to device}} \div$$

        = 22 500 J ÷ 112 500 J

        = 0.2 (or 20% if the second version of the equation was used)

> **e)** The student was only 20% efficient as she walked to the music shop.
> What happened to the rest of the 112 500 J that she used?

Example 2 A student using a 1200 W hair drier

Of the 1200 watts of power (1200 joules of energy every second)
suppose only 480 W are useful in giving hot fast moving air:

$$\text{Efficiency} = \frac{\text{power output}}{\text{power input}} \times 100\%$$

        = (480 W ÷ 1200 W) × 100%

        = 40% (or 0.4 if first version of equation used)

> **f)** Where do you think the remaining 720 W are wasted in the hair drier?

Remind yourself!

1 Copy and complete:

Energy ...... helps us to compare different
processes. The most efficient processes ......
less energy. They convert ...... of their input
energy into ...... output energy.

Equation 7

$$\text{......} = \frac{\text{...... energy transferred by device}}{\text{total energy ...... to device}} \div$$

This equation gives a decimal fraction value for
efficiency. Often the answer is given as a ......
by multiplying it by 100%.
It is i...... for something to have an efficiency
greater than 1 (or 100%)

2 Copy and complete this table:

Device	Input	Output	Efficiency (% or decimal)
Washing machine	2500 W	500 W	20% or 0.2
TV	250 W	50 W	
Toaster	1000 J		40% (0.4)
Computer	3000 J	1500 J	
Light bulb	100 W		5% (0.05)

3 Why is it impossible to have a machine or
process that is more than 100% efficient?
(Hint: Law of Conservation of Energy.)

Which hurts more when it hits your toe after you
drop it, a basketball or a ten pin bowling ball?
The bowling ball has more mass so it has more
gravitational potential energy while you are holding it.
This means that when you drop it, it gains more
kinetic energy as it falls.
Can you imagine trying to get a bowling ball through
a basket ball net!

Look at this list:

- A bullet travelling at 270 mph;
- A formula one car travelling at 180 mph;
- A scorpion tank travelling at 56 mph;
- You riding a bike, travelling at 16 mph;
- An oil tanker travelling at 4.5 mph.

These have been listed in order from fastest to slowest.

a) Which do you think would have the most kinetic energy?

Name	Velocity (mph)	Kinetic energy (J)
Bullet	270	72
Formula 1 car	180	2 560 000
Scorpion tank	56	9 900 000
You riding a bike	16	1764
An oil tanker	4.5	1 000 000 000

b) Arrange the kinetic energy values above into order from most to least.
Was your answer to a) correct?

Kinetic energy depends on **how fast** the object is **moving**.
It also depends on the **mass** of the object moving.
Normally when we think of how fast something is moving,
we talk about the object's speed.
Often in physics we use the term **velocity** instead of speed.
Velocity gives more information about a moving object.
It gives us the object's speed and also which way it is going
(its direction).

Example   Speed: a car travelling at 50 mph
Velocity: a car travelling at 50 mph *north*

(Speed and velocity are explained in more detail in Chapter 10.)

Calculating kinetic energy

The kinetic energy figures in the table on the last page were calculated using the kinetic energy equation:

Equation 8 to learn

> **Kinetic energy $= \frac{1}{2} \times$ mass $\times$ (velocity)^2**
>
> Symbol form: $KE = \frac{1}{2}mv^2$

In equation 8 we have (velocity)^2
We say velocity squared.
So if the velocity was 12 m/s,
(velocity)$^2 = 12 \times 12 = 144$ (m/s)^2

c) Work out the units for each part of the kinetic energy equation.

Example 1 You riding a bike

Calculating the amount of kinetic energy that you have when riding a bike. 16 mph is approximately 7 m/s
Your mass could be 60 kg
The bike's mass   12 kg
Total mass   72 kg

$$
\begin{aligned}
\text{Kinetic energy} &= \tfrac{1}{2} \times \text{mass} \times \text{(velocity)}^{2 \text{ (squared)}} \\
&= \tfrac{1}{2} \times 72 \text{ kg} \times (7 \text{ m/s})^2 \\
&= 36 \times 49 \\
&= 1764 \text{ J}
\end{aligned}
$$

d) Use the example above to help you work out the kinetic energy of you and the bike when you are travelling at:
   i) 5 m/s   ii) 10 m/s   iii) 15 m/s (down a steep hill)

Example 2 A ball moving

Look carefully at this table for a basket ball travelling at different velocities.

Type of ball	mass (kg)	$\frac{1}{2} \times$ Mass (m/s)	Velocity	(Velocity)^2	Kinetic energy (J)
Basket ball	1	0.5	1	1	0.5
Basket ball	1	0.5	2	4	2
Basket ball	1	0.5	4	16	8
Basket ball	1	0.5	8	64	32
Basket ball	1	0.5	16	256	128

Remind yourself!

1 Copy and complete:

The amount of kinetic energy that an object has depends on its ...... and how ...... it is moving. ...... has the same number value as ...... but has a direction as well.

Equation 8
...... energy $= \frac{1}{2} \times$ mass $\times$ (......)^2

2 Look at the table above in Example 2
In each new line the velocity of the basket ball has been doubled.
Have the kinetic energy values doubled?

3 Calculate the kinetic energies for a 2 kg bowling ball rolling at the following velocities; 1 m/s, 2 m/s, 4 m/s, 8 m/s, 16 m/s.

Summary

Equations to learn:

Equation 1

work done = energy transferred

Equation 2

work done	=	force applied	×	distance moved
or energy transferred				in the direction of the force
Units joule (J)		newton (N)		metre (m)

Equation 3

change in gravitational potential energy	=	weight	×	change in vertical height
Work done *upwards*, energy transferred *upwards*		a force		a distance moved *upwards*
Units joules (J)		newton (N)		metre (m)

Equation 4

$$\text{power} = \frac{\text{work done (joules, J)}}{\text{time taken (seconds, s)}}$$
(watt, W)

This equation is often written to calculate energy transfer instead:

energy transferred = power × time
(work done)

Equation 5 Electrical energy

energy transferred	=	power	×	time
Units kilowatt hour (kWh)		kilowatt (kW)	×	hours (h)

Equation 6 Cost of electrical units

total cost	=	number of electrical units	×	cost per unit
(£ and pence)		(number of kWh used)		(pence each)

Equation 7 Efficiency

$$\text{efficiency} = \frac{\text{useful energy transferred by device}}{\text{total energy supplied to device}}$$

This version of the efficiency equation gives efficiency as a decimal fraction (*always less than one*).
or

$$\text{efficiency} = \frac{\text{useful output energy}}{\text{total input energy}} \times 100\% \qquad \text{Efficiency} = \frac{\text{power output}}{\text{power input}} \times 100\%$$

These versions of the equation give efficiency values as a percentage (*always less than 100%*).

Equation 8

kinetic energy = ½ × mass × (velocity)² (squared)

Symbol form: $KE = \frac{1}{2}mv^2$

Questions

Key point:
Whenever doing a calculation, always write down the equation you are using first!

1 i) A table was dragged 5 m across a classroom floor using a force of 50 N. How much work was done?

ii) A student opened a classroom blind using a force of 16 N. If the total work done lifting the blind was 32 J, how far was the blind moved?

iii) A student has to push the sofa nearer to the TV to relax after a hard day at school. He moves it by 50 cm using 40 J of energy. What force did he use?

2 i) A student is told to put her bag in her bedroom, not leave it lying around downstairs. If the bag weighs 60 N and she has to carry it 4 m upstairs, how much gravitational potential energy does the bag gain?

ii) A first time parachutist weighing 700 N is just about to jump (get pushed really) for the first time. The aeroplane is level at 1000 m when they go, how much gravitational energy do they have as they jump?

iii) A student drops a box full of text books on their toe. The books had 120 J of gravitational potential energy before they dropped from a bench 120 cm high. How much did the books weigh?

3 The table below shows the results for a group of students measuring their power rating using bell-bars. Copy and complete the table, finding the most powerful student.

Name	Energy per lift	No. lifts	Total energy (J)	Time (s)	Power (W)
Josh	30	25	30 × 25 =	50	
Chris	24	25		60	
Leanne	25	16	400	40	
Steven	70	40		140	
Tammy	16	3	48	24	

4 i) James uses his sister Nicola's 1 kW hair drier and mousse to shape his hair. If he runs the hair drier for 15 min, how much energy has he transferred?

ii) Nicola decides to get James back by using his hi-fi on full blast while he is out. The hi-fi uses 0.5 kW and she has it on for 3 hours. How much energy was transferred by the hi-fi?

iii) James and Nicola's mum decides to have a hot bath to avoid the noise. She runs the immersion water heater for half an hour, replacing the water James used. If the heater converted 1 kWh of electrical energy into heat, what was its power?

5 For each of the following, find the peak and the off-peak answers. Use 8p per unit for peak electricity and 2p per unit for off-peak.

i) Calculate the cost of using a 100 W light bulb for 20 hours.

ii) How long would it take a 2 kW electric radiator to convert 32p of electricity into heat?

iii) Find the cost of electricity in each part of question **4** above.

6 Efficiency-based calculations:

i) Find the efficiency of an electric motor supplied with 200 J of energy that gives 50 J of useful work.

ii) A mobile phone charger has an efficiency of 65%. Calculate how much useful energy it transfers when supplied with an input of 100 J.

7 Which has more kinetic energy, an 800 kg motorbike travelling at 50 m/s, or a 20 000 kg truck travelling at 10 m/s?

1 The diagram shows an experimental solar-powered bike.

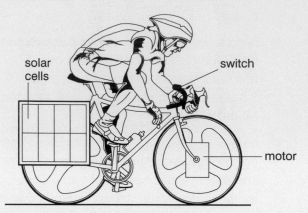

solar cells

switch

motor

A battery is connected to the solar cells.
The solar cells charge up the battery.
There is a switch on the handlebars.
When the switch is closed, the battery drives a motor attached to the front wheel.

(a) Rewrite the sentences. Use words from the list to complete them. Words may be used once, more than once, or not at all.

> **chemical   electrical   heat (thermal)**
> **kinetic   light   potential   sound**

  (i) The solar cells transfer …… energy to …… energy.

  (ii) When the battery is being charged up, …… energy is transferred to …… energy.

  (iii) The motor is designed to transfer …… energy to …… energy. (6)

(b) (i) You may find this equation useful when answering this part of the question.

$$\text{power (watt), W} = \frac{\text{work done (joule, J)}}{\text{time taken (second, s)}}$$

The cyclist stops pedalling for 10 seconds. During this time the motor transfers 1500 joules of energy. Calculate the power of the motor. (2)

  (ii) Name **one** form of wasted energy which is produced when the motor is running. (1)
(AQA (NEAB) 1999)

2 The table gives the percentage of electrical energy generated from different energy resources in Europe.

Energy resource	Percentage of electrical energy generated
Coal	50
Nuclear	20
Gas	15
Oil	10
Renewables	5

(a) Use data from the table to find each letter's label on the pie chart. One has been done for you.

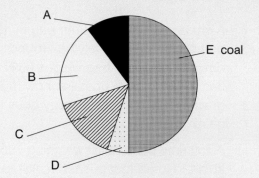

A

E coal

B

C

D

(4)

(b) Coal is a non-renewable energy resource.

  (i) What is meant by *non-renewable*? (1)

  (ii) Explain why wood is a *renewable* energy resource. (1)

  (iii) Name **two** other renewable energy resources. (2)
(AQA 2001)

3 Four groups of students investigate how the heat loss from a model house can be reduced. Each group has the same type of model house.

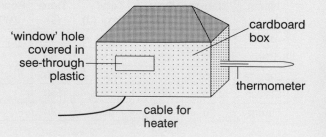

'window' hole covered in see-through plastic

cardboard box

thermometer

cable for heater

The 'house' is a cardboard box with a roof and hole for the window.
The hole is covered with a see-through plastic window.
An electric heater is used to warm the house.
A thermometer shows the temperature inside the house.

(a) The heater is switched on.
How does the thermometer show that the heater is transferring energy to the house? (1)
After twenty minutes, the heater is switched off.
The students record the temperature for another twenty minutes.

(b) Group 1 wrote this report.

GROUP 1
Our house was used as the 'control' for the other groups, so we did not change our house in any way.
The temperature fell when we turned the heater off.

(i) Why did the temperature fall when they turned the heater off? (1)

(ii) Why did this group not make any changes to their 'house'? (1)

(c) Each of the other three groups changed the basic model house in a different way.
Energy may be transferred by **conduction**, **convection** and **radiation**.
Use the words **conduction**, **convection** and **radiation** in your explanations of the students' observations.

(i) Group 2 wrote this report.

GROUP 2
We covered the roof of our house with silver foil.
The temperature did not fall as quickly as Group 1's house.

Explain the observations of Group 2. (2)

(ii) Group 3 wrote this report.

GROUP 3
We lined the inside of the roof of our house with cotton wool.
The temperature did not fall as quickly as Group 2's house.

Explain the observations of Group 3. (2)

(OCR 2001)

4 The diagram shows a section through a gas oven.

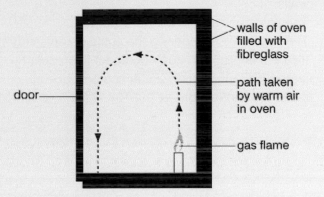

walls of oven filled with fibreglass

path taken by warm air in oven

door

gas flame

Rewrite the sentences. Use words from the list to complete them.

conduction   convection   insulation
radiation   resistance

The outside of the door gets hot because energy is transferred through the door by ....
Energy is transferred from the gas flame to the rest of the oven by the movement of air.
This type of energy transfer is called ......
The walls of the oven are packed with fibreglass to reduce energy transfer. Energy transfer is reduced because fibreglass provides good ......
The outside of the cooker is white and shiny.
This reduces energy transfer by ...... (4)

(AQA (NEAB) 1999)

5 The diagram shows a boy performing the standing long jump.

The boy weighs 600 N. In the middle of the jump, he has risen by 0.8 m. Use the equation:

potential energy = force × vertical height
       (J)   (N)   (m)

to calculate his potential energy when he has risen 0.8 m. (2)

(EDEXCEL 1999)

6 A leisure centre has security lights fitted on the walls.

The lights use 100 W filament lamps.

The council is comparing the costs of replacing the filament lamps with 20 W compact fluorescent lamps (CFLs). These CFL lamps produce the same amount of light as the 100 W filament lamps.

filament lamp   compact fluorescent
                                lamp (CFL)

The following table gives information about the two types of lamp.

Type of lamp	Lifetime of lamp (hours)	Power (kW)	Cost of 1 kWh of electricity (£)	Cost of electricity to use lamp for its lifetime (£)	Cost of one lamp (£)	Total cost of lamp and electricity (£)
filament	1 000	0.1	0.08	8.00	0.50	8.50
CFL	12 000	0.02	0.08	—	10.00	—

(a) One CFL lasts as long as 12 filament lamps. What is the cost of buying and using 12 filament lamps for their lifetime (1 000 hours each)? (2)

(b) The cost of electricity can be calculated from the equation:

$$\text{Cost of electricity} = \text{power (kW)} \times \text{time (h)} \times \text{cost of 1 kWh}$$

(i) What is the cost of the electricity used by a CFL lamp during its lifetime of 12 000 hours? (2)

(ii) What is the total cost of buying and using a CFL for 12 000 hours? (1)

(c) Use your answers and the information in the table to give TWO benefits of changing to CFLs. (2)

(EDEXCEL 1998)

7 The table shows the power rating of appliances used in the home.

Appliance	Power rating
fire	2000 W
television	250 W
kettle	2·5 kW
hair drier	1200 W
vacuum cleaner	800 W

(a) If all the appliances were used for the same length of time, state which one would

(i) use the least amount of electricity, (1)

(ii) cost the most to run. (1)

(b) Calculate

(i) the power of the fire in kW, (1)

(ii) the cost of using the fire for 2 h if one unit of energy (kWh) costs 8p.
You may use the following equations

$$\underset{\text{(kWh)}}{\text{Energy used}} = \underset{\text{(kW)}}{\text{power}} \times \underset{\text{(h)}}{\text{time}}$$

$$\underset{\text{of units}}{\text{Cost}} = \underset{}{\text{number}} \times \underset{\text{unit}}{\text{cost per}}$$
(2)

(WJEC 1998)

Section Two
Electricity

In this section you will find how we use electricity.
You will learn about voltage, current and resistance
and how to calculate the electrical power of machines.
You will see how every motor driving machine relies
on both magnetism and electric charge to work.
The final chapter will help you to understand both
the uses and dangers of mains electricity in your home.

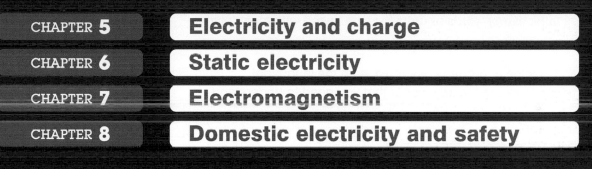

CHAPTER 5

Electricity and charge

▶▶▶ 5a Electrical definitions

Electricity is in every part of our lives. We use it hundreds
of times every day. Most people only think about how useful
it is when they have a power cut!
Suddenly every mains powered electrical device is useless.

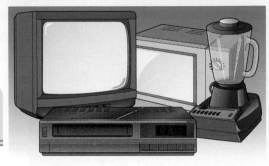

a) Write your own definition of electricity.

b) Can you think of any words or **units** that are linked with electricity?

To understand electricity properly, we must learn the language!
The 'terms' given here are the key words that you need to understand.

Electrical appliances.

Charge

> The unit of charge is the **coulomb (C)**
> The two types of electric charge are
> **positive** (+**ve**) and **negative** (−**ve**)

Both types of charge exist in every atom.
Negative charges repel other negative charges.
Positive charges repel other positive charges.
Positive charges like negative charges.
(We all know that opposites attract!)

> Like charges repel, unlike charges attract!

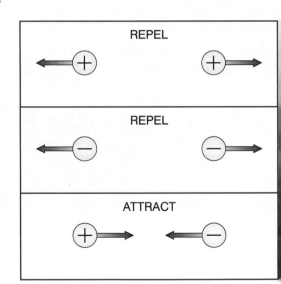

Current electricity

> The unit of electric current is the **ampere** (**A** or **Amps**).

When charges are able to move, we can get a **flow of charge**.
This is called an **electric current**. In metals, some of the negatively
charged electrons are free to move around. In fluids, electrons
and any **ions** (charged atoms and molecules) present can also
move and help a current to flow.

c) What do you think makes the charged particles move in one direction
in an electric current?

Static electricity or static charge

A TV screen often becomes 'charged' up. If you place your hand near its surface you can feel a 'tingle' and sometimes you will get a slight shock. This is because some **static** charge has collected on the screen. It can only flow away easily when a conductor touches it – through your hand in this case. See Chapter 6 for more detail.

Charge that is stationary (cannot flow) is called **static electricity**.

Voltage

Metals are full of free electrons (see page 37).
Liquids and gases can contain ions.
Just because they are there, it does not mean that an electric current will flow.

To make charge flow, it needs to be attracted to one point and repelled from another. For this to happen, the two points would need a **potential difference** between them (often called a **voltage**).

A potential difference or voltage between two points makes one of the points negative and the other point positive.
A charge that is free to move will travel to one of the two points.
If the charge is positive, it will be attracted and flow to the negative point.
If the charge is negative, it will be attracted and flow to the positive point.

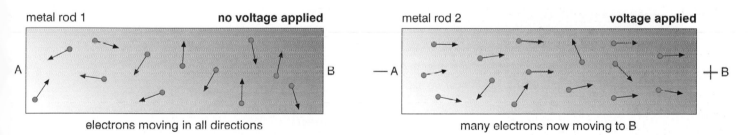

metal rod 1   **no voltage applied**

A ... B

electrons moving in all directions

metal rod 2   **voltage applied**

– A ... +B

many electrons now moving to B

The unit of potential difference (or voltage) is the **volt (V)**.

Remind yourself!

1 Copy and complete:

There are ...... types of electric charge, ... and ......
Like charges ......, unlike charges ...... electricity involves moving charge.
Static ...... is stationary charge.
Charges free to ...... will make an electric ...... if enough of them move in the same direction.
This will only happen if a ...... or potential ...... is applied.

2 Look at the diagram above.
Explain how the electrons are moving in each rod.

3 Find out about ions

i)   What are they?

ii) How their movement in a liquid allows a current to flow.

Think about these questions:
Which has more energy stored in it;
   a 12 V car battery or a 16 V cordless hair drier battery?
Which has the greatest current passing through it;
   an electric welding machine or an electricity pylon?

a) With a partner, discuss the two questions above and see
   if you can agree on your answers.

Current and voltage

A potential difference (or voltage) connected across a conductor will make
the electrons inside it flow. They will move towards the higher potential
(more positive) end of the conductor. This is because electrons
are **negatively** charged and are **repelled** from the negative end
of the conductor. They are **attracted** towards the positive end.
This is an example of '**real**' electric current.

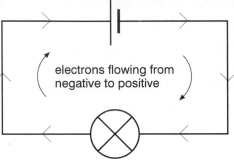

Real current.

Real current flows from **negative** to **positive**.

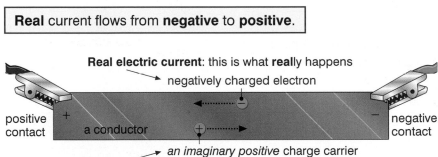

Real electric current: this is what **real**ly happens
negatively charged electron

positive contact   a conductor   negative contact

an *imaginary positive* charge carrier
Conventional electric current: this is how we explain current

Conventional current was defined by scientists before anyone knew
about the charge on electrons. *They got it wrong!*
It is still used to explain circuits.

Conventional current flows from **positive** to **negative**.

We use the colour codes red for positive and black (or sometimes blue)
for negative. So when you construct a circuit, the connections must
be from red to black (positive to negative).
Just make sure that you remember the real story;
the current really flows the other way.

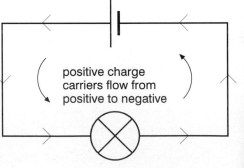

Conventional current.

b) Use your own words to explain the concepts of real
   and conventional current.

c) If the voltage does not flow, why do batteries get run down?

d) Is air a good conductor of electricity?

Electrical energy and circuits

A new battery is charged up. It has a higher potential at its positive terminal than at its negative terminal. It acts as an **energy store**. While the battery is not connected to a circuit, no charge can flow out of it. This means that the battery should keep its stored energy until it's required.

When the battery is fitted into a mini disc player, the stored energy inside it can be released to provide you with music. When the player is turned on, the potential difference between the battery's terminals makes electrons leave the negative terminal. All the electrons in the circuit between the two terminals of the battery start to move as well. *The electrons pass energy to the different components of the disc player as they pass through them.*

A range of batteries and cells.

e) What happens if the player is used for a long time?

Eventually so many electrons will have gone around the circuit to the positive terminal that its higher potential will start to decrease. The battery gets '**run down**' as the chemicals inside it are used up. If you have the volume up louder, more energy is required. This means more charge needs to flow around the circuit. A larger current makes the battery run down a little quicker.

Electrical power

Power is the amount of energy transferred every second. Equation 4 on page 50 is used in other energy systems. In electrical circuits we can calculate power using equation 9.

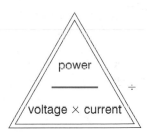

Equation 9 to learn

	power = voltage × current		
Units	watts	volts	amps
	(W)	(V)	(A)

Example Power rating of a toaster

Mains voltage = 230 V   Current = 5 A   power = voltage × current
= 230 V × 5 A = 1150 W

Remind yourself!

1 Copy and complete:

...... current involves electrons flowing from ...... to ...... This is from lower to ...... electrical potential. Conventional current is always used in ...... It flows from ...... to ...... When ...... flows through a circuit, ... is transferred to components of the circuit.

Equation 9
power = ... × ......

2 Copy and complete this table:

Appliance	Voltage	Current	Power
Kettle	230 V	6 A	
Headlamp	12 V		60 W
Light bulb		0.5	3 W

3 Find out about the work of Georg Ohm.

▶▶▶ 5c Resistance

Have you ever wondered why batteries seem to last longer in some things than they do in others?

- Put two 1.5 V batteries into a small torch. Leave it on all night and the batteries are 'dead' the next morning.
- Put two more identical 1.5 V batteries into a calculator and leave it on for a month. The chances are it will still help you to do your maths homework.

a) Why do you think the calculator's batteries last longer than the torch's?

Both the torch and the calculator have the same potential difference across their internal circuits. Two 1.5 V batteries wired in **series** (one after the other) gives a total voltage of 3 V. The torch must allow the charge to flow from its batteries at a faster rate than the calculator. How can a torch control the amount of charge that flows from its batteries?

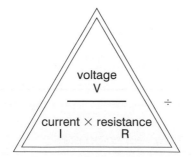

The circuits inside the torch have a much lower **resistance** so more charge can flow through them in a given time. The energy from the batteries is passed to the bulb to provide light and heat. *The only time the calculator gets hot is during a maths exam.*

> Resistance '**holds back**' the flow of electricity.
> The higher the resistance, the harder it is for charge to flow.

In any circuit, the size of the electric current (the amount of charge flowing per second) depends on the voltage applied and also the resistance of the circuit. **Georg Ohm** was the first to define this relationship between current, voltage and resistance. Both the unit of resistance and an equation are named after him!

Equation 10 to learn

Ohm's law				
(voltage)	**potential difference**	=	**current** ×	**resistance**
Symbol form	V	=	I ×	R
Units	volts (V)		ampere (A)	ohm (Ω)

Example What is the resistance of a car headlamp bulb?

A headlamp bulb in a 12 V car system will have a current of 5 A flowing through it.
From equation 10 (using the triangle to help rearrange)

potential difference ÷ current = resistance
$$12\,V \div 5\,A = 2.4\,\Omega$$

The factors that affect resistance

Have you ever walked around school during lesson time?
The place is deserted! You can walk through a narrow corridor or down
a tight staircase without fighting past hundreds of other students.
The resistance to your movement is much lower than normal!

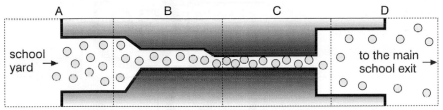

in this short-cut home, there are ten students in each section

The sketch above shows a short cut out of school.

b) In which section is the least resistance to student flow?

c) Where is the greatest resistance to student flow?

Factor 1   cross-sectional area

side views   cross-section views

Wire A

Wire B

Which wire has the greatest resistance to
the flow of current?

Wire B,
  it has a much smaller cross-sectional area.

Factor 2   length

Wire C

Wire D

these wires are the same type and diameter

Which wire has the greatest resistance to current
flow?

Wire D,
  it is much longer than wire C

Factor 3   type of material
Metals are good conductors.
They have low electrical resistance.
Rubber and plastic are good insulators.
They have high electrical resistance.

Factor 4   temperature
When the temperature of a metal
(and many non-metals) increases,
its electrical resistance also increases.

---

d) Are materials with a low electrical resistance good or bad conductors of heat?

Remind yourself!

1 Copy and complete:

The ...... through a conductor depends on the
...... across it and its resistance.
Resistance ...... the flow of charge.

Equation 10 Ohm's law
potential difference = ... × ......
In symbol form ...... = I × ......
The unit of resistance is the ...... (...)

2 Use Ohm's law to calculate:

i)   The voltage across a 4 Ω resistor which has a
     current of 3 A flowing through it.

ii) The current flowing through a 10 Ω resistor
     when 5 V is applied across it.

3 Make up a chart or poster to explain clearly the
    four factors that affect the resistance of a
    material.

▶▶▶ 5d Electrical circuit symbols

We spend our lives looking at symbols.
We use them to communicate information quickly.
What do the symbols on the right mean?

a) Can you think of another 5 everyday symbols?

In electrical circuits we have a set of symbols that
identify all the different components within a circuit.
You will have used them before, both in science and in CDT.

b) Draw as many electrical symbols as you can. (Close this book now!)

The main circuit symbols

Cell Cells are joined together to make a battery. In everyday life, people refer to cells as batteries. Domestic 1.5 V batteries are cells. *(Note: the longer line is the positive contact)*		**Battery** Batteries are made up of two or more cells. A 12 V car battery has six 2 V cells.	
Open switch 'Open' so that electricity can't flow around the circuit.		**Closed switch** 'Closed'. Now electricity can flow around the circuit (if the rest of it is complete).	
Lamp (or test bulb) This symbol is the most common one used for a bulb.		**Lamp** (or bulb) This symbol for a lamp is sometimes used	
Resistor Normally the resistance is written below the resistor.	5 Ω	**Variable resistor** These resistors do not have fixed values. Their resistance can be changed easily.	

c) Can you think of any common household uses of a variable resistor?
(Hint: lights are too bright. Stereo is too loud.)

d) How many cells do you think a 24 V lorry battery contains?

e) How many cells do you think a 6 V motor bike battery contains?

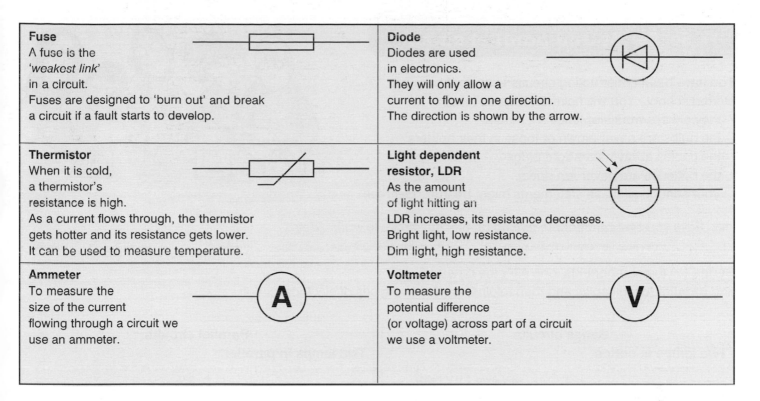

Fuse
A fuse is the 'weakest link' in a circuit.
Fuses are designed to 'burn out' and break a circuit if a fault starts to develop.

Diode
Diodes are used in electronics.
They will only allow a current to flow in one direction.
The direction is shown by the arrow.

Thermistor
When it is cold, a thermistor's resistance is high.
As a current flows through, the thermistor gets hotter and its resistance gets lower.
It can be used to measure temperature.

Light dependent resistor, LDR
As the amount of light hitting an LDR increases, its resistance decreases.
Bright light, low resistance.
Dim light, high resistance.

Ammeter
To measure the size of the current flowing through a circuit we use an ammeter.

Voltmeter
To measure the potential difference (or voltage) across part of a circuit we use a voltmeter.

Circuit diagrams

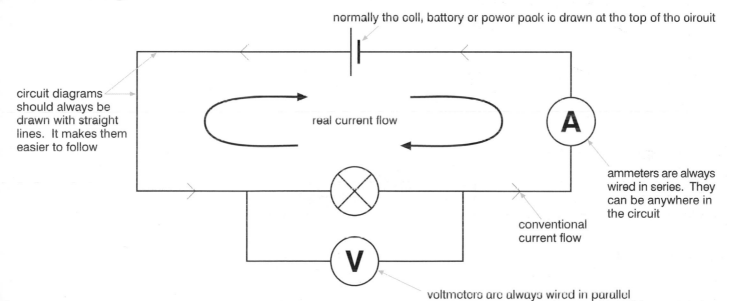

normally the coll, battery or power pack is drawn at the top of the circuit

circuit diagrams should always be drawn with straight lines. It makes them easier to follow

real current flow

ammeters are always wired in series. They can be anywhere in the circuit

conventional current flow

voltmeters are always wired in parallel

Remind yourself!

1 Copy and complete:

Electrical circuit …… are drawn using …… to represent the different components (parts) in the circuit.
They are drawn using …… lines.
The ……, batteries or power packs are normally placed at the …… of the diagram.

2 Find out the symbols for
 i)   a power pack.
 ii) an LED, light emitting diode.

3 In the circuit diagram above, the voltmeter is drawn in parallel and the ammeter is in series. What do the words **parallel** and **series** mean in circuits?

►►► 5e Series and parallel circuits

You have been constructing circuits ever since you first started school. You will have discovered that:
- they always work first time
- the bulbs are never blown or loose in their holders
- the cables always have tight plugs
- the batteries are never run down.

In your school, all these statements might be true. *Get real!*

a) Have you ever had trouble getting an electric circuit to work? Why?

Series (all the components, one after the other)
and **parallel** (circuit splits and then rejoins) are the two types of electrical circuit.

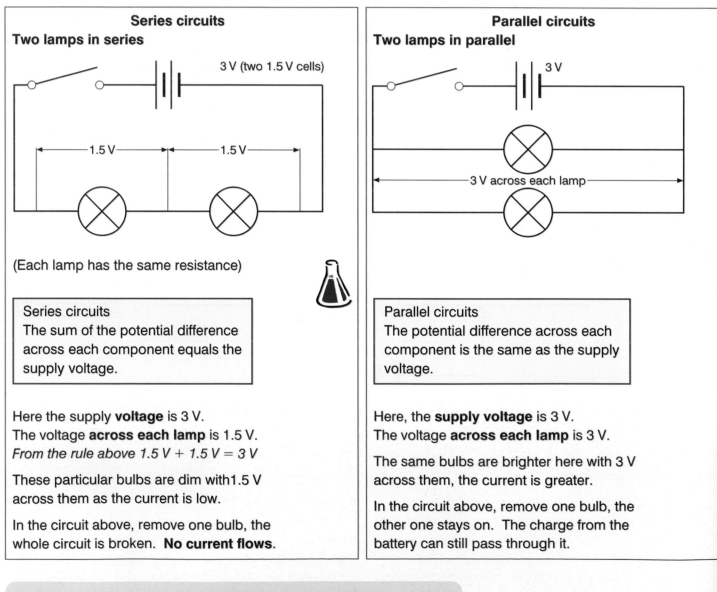

Series circuits
Two lamps in series

3 V (two 1.5 V cells)

1.5 V — 1.5 V

(Each lamp has the same resistance)

> **Series circuits**
> The sum of the potential difference across each component equals the supply voltage.

Here the supply **voltage** is 3 V.
The voltage **across each lamp** is 1.5 V.
From the rule above 1.5 V + 1.5 V = 3 V

These particular bulbs are dim with 1.5 V across them as the current is low.

In the circuit above, remove one bulb, the whole circuit is broken. **No current flows**.

Parallel circuits
Two lamps in parallel

3 V

3 V across each lamp

> **Parallel circuits**
> The potential difference across each component is the same as the supply voltage.

Here, the **supply voltage** is 3 V.
The voltage **across each lamp** is 3 V.

The same bulbs are brighter here with 3 V across them, the current is greater.

In the circuit above, remove one bulb, the other one stays on. The charge from the battery can still pass through it.

b) Draw a series circuit with two cells. Include a variable resistor controlling the brightness of one bulb.

c) What would be the total potential difference in a circuit when three 2 V cells are connected together in series.

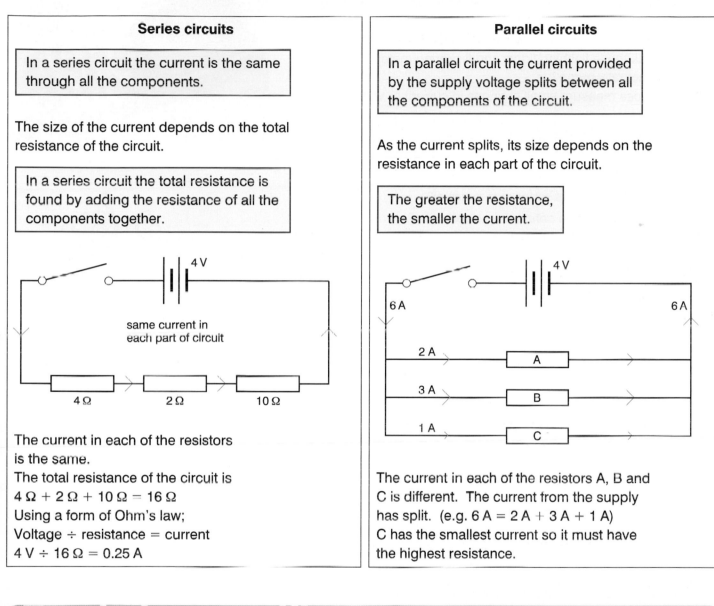

Series circuits

In a series circuit the current is the same through all the components.

The size of the current depends on the total resistance of the circuit.

In a series circuit the total resistance is found by adding the resistance of all the components together.

4 V

same current in each part of circuit

4 Ω   2 Ω   10 Ω

The current in each of the resistors is the same.
The total resistance of the circuit is
4 Ω + 2 Ω + 10 Ω = 16 Ω
Using a form of Ohm's law;
Voltage ÷ resistance = current
4 V ÷ 16 Ω = 0.25 A

Parallel circuits

In a parallel circuit the current provided by the supply voltage splits between all the components of the circuit.

As the current splits, its size depends on the resistance in each part of the circuit.

The greater the resistance, the smaller the current.

4 V

6 A   6 A

2 A   A

3 A   B

1 A   C

The current in each of the resistors A, B and C is different. The current from the supply has split. (e.g. 6 A = 2 A + 3 A + 1 A)
C has the smallest current so it must have the highest resistance.

Remind yourself!

1 Copy and complete:

The total ...... of a series circuit is calculated by ...... up all the resistances in the circuit.
The ...... provided by the cell or battery is shared between all the components in series.
In a ...... circuit, the potential difference is the ...... across each component.
The current splits between the different parts of the circuit, the components with the ...... resistance having the greatest ......

2 In a series circuit, a 3 Ω, a 6 Ω and a 9 Ω resistor are connected together.

The total supply voltage is 6 V.

i)   Calculate the current in the circuit.

ii) What is the voltage across each of the resistors?

3 If the resistors in question **2** were connected in parallel, which would have the greatest current flowing through it?

Think about these questions:
- How does your mobile phone charger know when the battery is fully charged?
- Have you ever wondered how a modern car measures its speed?
- What do all those flashing lights and flickering meters on stereo systems actually mean?

All these devices measure small changes in current or voltage.

We use ammeters to measure current flowing through a circuit. Voltmeters measure the voltage (potential difference) between two points in a circuit. *The 'voltage drop' across a component in a circuit.*

a) Are ammeters wired in series or in parallel?

b) Which way would you connect up a voltmeter?

A stream of water

You are asked to measure the current in a stream of water.
Will you:

look at the stream twice, 5 m apart and try to spot a difference;

or

jump straight in and feel the 'strength' of the current of water flowing past!

To measure the size of current, you have to be **in the flow** of the current.

Ammeters measure the size of electric current by being **wired in series**.
The current has to pass directly through them. (They have a very low resistance.)

A log ride

You are at a theme park with two water log rides.
To find the one with the greater change in gravitational potential of water would you:

stand at the bottom to see which one splashed you the most

or

work out which one has the largest drop.

To measure a **change in potential**, you need to measure the potential both before and afterwards.

Voltmeters are wired **in parallel**. This means that they measure the electrical potential difference between two points in a circuit.

---

c) What do you think would happen to a two bulb series circuit if you put a voltmeter in series with the bulbs?

Cells and batteries in circuits

How many times have you put batteries into a CD player
or a portable radio?

d) How are the batteries connected inside a torch, in series or in parallel?

Most small torches use at least two batteries wired up in series.
So what happens to the total potential difference when you wire
up cells or batteries in series?

> The total potential difference of
> cells connected in series is the
> sum of all their voltages.

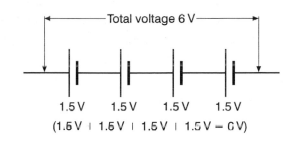

Total voltage 6 V

1.5 V   1.5 V   1.5 V   1.5 V

(1.5 V + 1.5 V + 1.5 V + 1.5 V = 6 V)

e) What would be the total potential difference in the circuit above if one of
the batteries was connected the other way round?

Batteries or cells are not always connected in series.
If a car battery runs down, 'jump leads' are often used to
help it to start.
The leads are connected positive terminal to positive,
negative to negative.
This means that the two car's batteries are connected
in parallel.

> For batteries with the same potential difference
> connected in parallel, the total potential
> difference stays the same as each of the batteries.

Together they take much longer to drain down
as they store more energy.

f) What do you think would happen if you connected
a 5 V battery in parallel with a 10 V one?

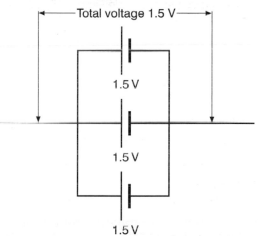

Total voltage 1.5 V

1.5 V

1.5 V

1.5 V

It is quite likely that you have built some circuits using both voltmeters and ammeters. *You might even have made the circuits work first time by connecting the meters in properly.*

Once you have measured the current through a bulb and the voltage across it a few times you are an expert! Here are some sample results for a 12 V filament lamp:

Voltage V	0	2	4	6	8	10	12
Current A	0	0.2	0.4	0.55	0.65	0.7	0.72

a) Draw a line graph of the results in the table above.
   Plot the voltage values on the *x* axis (along the bottom) and current values on the *y* axis (up the side).
   Look at your graph for **a)**

b) Does increasing voltage increase the current?

c) What happens at the top of the graph?

Resistor at constant temperature

If you **increase the voltage** across a resistor, the **current will increase**.

On the negative side of the graph, the only difference is that the voltage is applied to the resistor the other way round. (The terminals are reversed). This makes the current flow in the opposite direction. The relationship between voltage and current is still true. When the voltage is increased, the current increases.

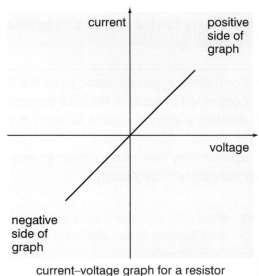

current–voltage graph for a resistor at constant temperature

In a resistor at **constant temperature**, increasing the voltage gives a steady increase in the current.

d) What do you think happens to the line if the temperature of the resistor is allowed to increase? (See factor 4 page 69.)

Filament lamps

Filament lamps are used everywhere.
They are the bulbs with a tiny piece of thin wire inside.
The current–voltage graph for a filament lamp
is drawn on the right.
*(Your graph answer to question a) should be
similar to the positive side of this graph.)*

On the positive side of the graph, the current
increases at first and then levels out. The voltage
is increased by even amounts but the current
does not increase evenly. On the negative side
of the graph, exactly the same happens.

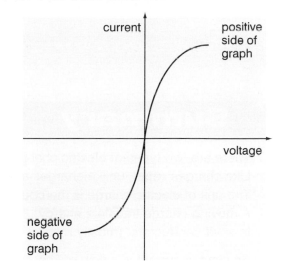

current–voltage graph for a filament lamp

> In a filament lamp, as the voltage gets higher, the current reaches a maximum value.
> The current levels out.

Diode

Diodes are used in many electronic circuits.
They are very useful because they stop the
current from flowing in one direction.
The direction that they allow electricity to flow in
is shown by the direction of the arrow in the symbol.

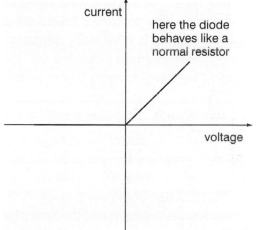

No current is flowing here. The diode is connected
so that the current is trying to flow against the arrow.

e) Sketch the graph you would get if you put two diodes
next to each other, but one reversed?

Remind yourself!

1 Copy and complete:

Current–...... graphs represent how the ......
through a device changes when different
voltages are applied across it.
The graph for a ...... lamp levels off because a
greater current produces heat which increases
the lamp's ......
A resistor kept at a ...... temperature has a
straight line graph. So an increase in voltage
gives an even increase in current.

The graph for a ...... only goes up in the
direction that current is allowed to ......

2 Which would have the steeper line if drawn on
the same axis, a 4 Ω resistor or an 8 Ω resistor?

3 When current is plotted on the x axis and voltage
on the y axis, the gradient of the graph is the
resistance. Plot these values this way round and
then find the gradient (resistance).

Current (A)	0	0.25	0.5	0.75	1	1.25	1.5
Voltage (V)	0	1	2	3	4	5	6

Summary

There are two types of electric charge; **positive** and **negative**.
Like charges **repel**, unlike charges **attract**.
The unit of electric charge is the **coulomb (C)**.
A moving charge transfers energy.
In solid conductors, (mostly metals) the charge carrier is the **electron**.

An electric current is a flow of charge.
The unit of electric current is the **ampere** or **amp (A)**.
One amp is a charge of one coulomb flowing per second.

For current to flow through a conductor, a potential difference (or voltage) must be applied across the conductor.
The unit of potential difference (or voltage) is the **Volt (V)**.

The system of electricity the world uses is called '**conventional current**':
a positive charge carrier flows from positive to negative.
Real current is what *really* happens:
negatively charged electrons are repelled by lower potential (negative) and they travel towards higher potential (positive).

Equation 9   Power in electrical circuits

$$\text{power} = \text{voltage} \times \text{current}$$
$$P = V \times I$$

| Units | watts (W) | volts (V) | amps (A) |

Resistance 'holds back' the flow of electricity.
The higher the resistance, the more the flow of charge is opposed.

Equation 10   Ohm's law

$$(\text{voltage}) \text{ potential difference} = \text{current} \times \text{resistance}$$

| Units | volts (V) | amps (A) | ohm (Ω) |

The four factors that affect resistance are:
cross-sectional area, length, type of material and temperature.

Properties of series circuits	Properties of parallel circuits
Broken circuit, no current flows anywhere.	Broken circuit in parallel branch, current still flows elsewhere.
Total resistance is the sum of resistors.	Current through part of circuit depends on resistance in that part of the circuit.
Voltage shared between each component.	
Cells or batteries connected in series, the total voltage is the sum of the individual voltages.	Voltage across each branch is the same as the supply voltage.
Ammeters are always wired in series.	Voltmeters are always wired in parallel.

Questions

1 Copy and complete:

i) There are two types of electric charge, ...
and ...... Similar or ...... charges repel,
unlike charges ......
The unit of electric charge is the ...... (C)
Moving charges ...... energy.

ii) Current is a flow of ...... and is measured
in ......

iii) A voltage (or a ...... difference) must be
present across a ...... for a current to flow
through it.
Voltage is measured in ......

iv) Conventional current is used in circuits.
The current flows from ...... to negative.
Real current is how charge actually flows.
Electrons flow from ...... to positive.

v) Electrical power is calculated using:
Power = voltage × ...... (P = ... × I)

vi) Resistance *holds back* the flow of ......
Resistance is measured in ......
Ohm's law: Voltage = current × ......
The four factors that affect resistance are;
cross-sectional area, ..., type of ...... and
t......

2 i) A current of 2 A passes through a TV
connected to a mains supply of 230 V.
Calculate the power of the TV.

ii) What would be the current used by a
1500 W, 230 V mains fan heater?

iii) Calculate the power of a 12 V headlamp bulb
that has a current of 5 A passing through it.

3 Use Ohm's law to calculate:

i) The voltage across an 8 Ω resistor which
has a current of 2 A flowing through it.

ii) The current flowing through a 200 Ω resistor
when 4 V is applied across it.

iii) The resistance of a bulb that has 6 V across
it and 0.5 A flowing through it.

4 Draw circuit diagrams for the following:

i) A series circuit with two bulbs, a battery of
two cells, an ammeter and a switch.

ii) A parallel circuit with two bulbs in parallel
with a resistor. Include a voltmeter across a
battery of three cells.

iii) A dimmer switch type of circuit using a
variable resistor to adjust the brightness of
a single bulb.

iv) A parallel circuit with two bulbs. Include a
diode so that one of the bulbs will only light
when the battery is fitted one way. Both
bulbs should light when the battery is fitted
the opposite way around.

5 Invent a simple saying to help you to remember
the correct way of connecting voltmeters and
ammeters into circuits.

6 Sketch from memory (without checking back)
the voltage–current graphs for:

i) A filament lamp.

ii) A resistor.

III) A diode.

7 Make up a small set of cards to help you learn
the following:

i) The electrical power equation,

ii) Ohm's law,

iii) The units for voltage, current, charge and
resistance;

iv) The electrical circuit symbols.

Example:

CHAPTER 6

STATIC ELECTRICITY

▶▶▶ 6a Stationary and moving charges

Have you ever combed your hair and made it stand up without using a spray or gel?

> Static electricity is charge that is '*stuck*'. It is stationary on the surface of an object and can't flow away.

Science text books which include electricity always have the 'picture of girl with long hair'. Her hair is standing on end and she is holding the top of the Van de Graaff generator.

a) Why does her hair spread apart and stand on end?

Have you seen the Van de Graaff generator used for any other 'tricks'? To explain clearly how objects get charged up, we need to understand exactly what charge is and how it moves.

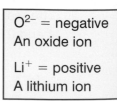

Isn't this normally a student's job?

The atom and charge

Everything is made from very tiny particles called atoms. There are three even smaller particles that form atoms. These are the **neutron**, the **proton** and the **electron**.

> What electric charge does each particle have?
> **b)** neutron **c)** proton **d)** electron

Atoms are **neutral**. They have no overall charge. So a neutral atom has the same number of protons and electrons.
An atom of the metal beryllium will have 5 neutrons, 4 protons and 4 electrons.
Neutrons have no electric charge.
The positive charge of the protons cancels out the negative charge of the electrons.
Ions are atoms (and molecules) that are charged.
A positively charged ion has more protons than electrons.
It will have lost some of its electrons, making it more positive.
A negatively charged ion has more electrons than protons.
It will have gained some electrons from another atom.

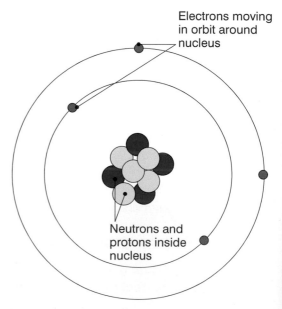

Electrons moving in orbit around nucleus

Neutrons and protons inside nucleus

A sketch of a beryllium atom.

> O^{2-} = negative
> An oxide ion
>
> Li^+ = positive
> A lithium ion

Stationary and moving charges

Like charges repel, unlike charges attract.

In the last chapter we saw that electrical conductors
have free electrons which move from lower to higher potential
(negative to positive). This follows the rule:
The negatively charged electrons repel the negative.
They are attracted by the positive and move in that direction.

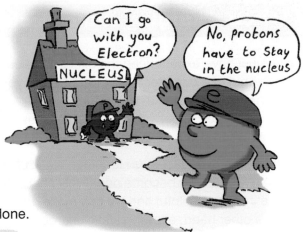

Electrons are the only charge carriers that can move in solids.

Protons are held inside the nucleus of an atom and rarely move alone.

e) Hydrogen has no neutrons, one electron and one proton.
When hydrogen forms an ion, it loses its electron.
What is a hydrogen ion really, and what is its charge?

A solid insulator has no free electrons. No charge can travel through it.
Any charge that collects on its surface is stuck there. It is **static charge**!

In conducting liquids, the charge carriers are ions (as well as the electrons
in molten metals). A charge can pass through a liquid if the liquid contains ions.
A potential difference from a pair of **electrodes** (electrical contacts)
would be needed. You dip the electrodes in the liquid.

The positive ions would move to the lower potential electrode. (−)
The negative ions would move to the higher potential electrode. (+)

(See page 89 for more detail about currents in liquids.)

Charge moving in gases also involves ions and electrons.
Gases are normally good insulators. They don't have many charge carriers.
If the potential difference is high enough, more ions can be formed.
This allows sparks to jump suddenly through a gas.
This is exactly what happens when lightning strikes.
You have probably seen sparks jumping from a Van de Graaff generator.

Remind yourself!

1 Copy and complete:

Atoms contain neutrally charged …… and ……
charged protons inside a central nucleus. They
have negatively charged …… in orbit around
their nucleus.
Uncharged atoms have the …… number of
protons and electrons.
…… ions are atoms that have lost electrons,
negative ions have …… electrons.
In conductors …… electrons move.
In insulators, no …… can move. In fluids, ……
and electrons move.

2 No charge can move on or near the surface of an
insulator.
If you rub a strip of nylon, it becomes charged.
Try to explain how.

3 Using a chemistry text book or a
computer to help, find out what type
of ions:

i) Group I metals form.

ii) Group II metals form.

iii) Group VII non-metals form.

You must have experienced the effects of releasing static charge:

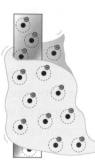

- Get out of a car after a long journey and you can feel a 'shock' when your feet touch the ground.
- Some types of polyester or nylon clothes can charge up your hair when you take them off. You can even hear a spark crack.

Whenever two surfaces rub together, static charge can build up.
One surface would become negatively charged and the other positively charged.
Sometimes it is helpful and we can make good use of the static charge.
On the other hand, allowing static charge to build up is very dangerous and could cause an accident if it is not discharged immediately.

a) Try to list some more examples of static charge collecting on a surface.

b) Can you think of any places where high static charge is very dangerous?

Next time you catch a plane or travel in a bus, look out for a rubber strap trailing on the ground. These straps **discharge** static electricity from vehicles when they stop moving. This stops sparks jumping to or from the Earth.

c) Why is the winch line from a rescue helicopter allowed to drag on the ground before the rescuer reaches the ground?

winch line

Build-up of charge is common and can be very dangerous.
Knowing how charge can build up on a surface is helpful.
Knowing how to make a highly charged object safe could be life saving.
Sparks near fuel can cause explosions!

Charging and discharging

When two surfaces rub together, electrons are swapped.
They leave the atoms on one of the surfaces and they join the other surface.

Charging:
The surface that has **lost electrons** becomes **positively charged**.
The surface that has **gained extra electrons** becomes **negatively charged**.

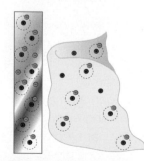

Before the cloth is rubbed on the rod (neutral charge).

To discharge a charged object, especially a Jumbo jet just coming in to land, connect it to Earth. A cable or strap touching the ground allows electrons to flow, neutralising the charged object.

Discharging to Earth:
If the object is positively charged, the electrons flow from the ground to it.
If the object is negatively charged, the electrons flow from it to the ground.

After the cloth has been rubbed on the rod (nylon is negative, cotton is positive).

Charged and neutral surfaces

The atoms on the surface of an insulator, like all atoms, have negatively charged electrons orbiting around a positively charged nucleus. When you move a negatively charged object close to the surface of a neutral insulator, things happen!
The electrons in atoms at the surface are repelled and want to escape. They can't get away so they 'hide' behind the positively charged nucleus. This leaves the top surface of the insulator positively charged, attracting the negative object.

strip of nylon

protons attracted to electrons

neutral solid

d) Describe what would happen when a positively charged object was brought close to a neutral one.

Charging up a metal object

Metals are conductors, how can they be charged up?
The electrons inside a metal will always flow towards a higher potential (+).
Connect a metal sphere to a positively charged rod and electrons will travel from the sphere to the rod.
Soon the sphere will have less negative charge.
Removing the rod will leave the sphere positively charged.

Step 1

Step 2

Electrons move to rod from sphere

Rod now neutral, the sphere is left positively charged

e) Describe a process similar to the one above where a piece of negatively charged nylon could be used to charge a neutral sphere.

When positive meets negative

When a positively charged acetate rod is brought near to a negatively charged nylon rod, they are attracted.
What if they are allowed to touch?
The extra loose electrons on the surface of the nylon rod move across to the acetate rod. They combine with the atoms that lost electrons when the acetate was originally charged up.
Both the nylon and the acetate rod becomes neutral.

Step 1

Electrons move across to positive rod

Step 2

Both rods now have neutral charge

Remind yourself!

1 Copy and complete:

Rub two different materials together, they can become ...... up.
Rub a piece of nylon with some cotton and ...... from the cotton will move across to the nylon making it ...... charged.
If you rub acetate with cotton, it becomes ...... charged, electrons are rubbed off its surface.
Connecting a charged object to ...... allows electrons to flow, discharging the object.

2 Use the idea of charged and neutral surfaces, described above, to explain how cling film pulled off the roll can stick to glass and china so easily. (Hint: pulling the cling film off the roll charges it up.)

3 Why is it best not to use an umbrella or hide under a tree during a thunder and lightning storm?

83

The last time you were in ICT, you probably printed some work out.
How does the printer work?
When did you last use a sheet of photocopied work?
How does a photocopier copy?
You might have your own bike or your family probably has a car.
How is the paint on bikes and cars made so even?

Inkjet and laser printers, photocopiers and commercial paint spraying
all use static electricity. There are many industrial processes
and a wide variety of machines that work because of static electricity.
In the next few pages we shall look at some of these applications.

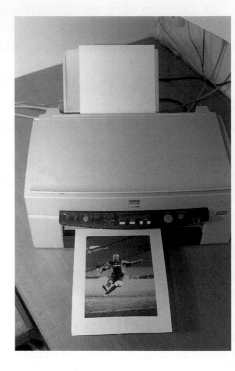

> **a)** Can you think of any domestic appliances which have a 'perfectly even'
> coat of paint?
>
> **b)** Can you work out how charging paint makes it spray so evenly?

Inkjet printers

One of the most common types of printer used
in offices, schools and at home is the inkjet printer.
They are cheaper than laser printers, reasonably
fast and normally give a good quality print-out.
Inkjet printers use static electricity to focus tiny
droplets of ink onto the paper in exactly the
right place to give clear characters.
The diagram below outlines how they work.

2. High potential difference between positive and negative plates.

paper

1. Ink in nozzle is charged and then pushed out in tiny droplets.

The

+

−

4. The ink droplets hit the page, making tiny dots. As more
ink collects characters begin to be formed on the page.
The voltage between the plates is then adjusted to focus
the inkjet on to a new part of the page. More characters
can then be drawn.

3. Charged droplets are repelled from
one plate and attracted towards the
other plate.
This makes the path of the ink
droplets bend as they are focussed
towards one point on the page.

> **c)** If the potential difference between the plates in the sketch above was
> increased, how would the path of the next ink droplet change?

Paint spraying

The method used to give an even coat of spray paint is similar to the one used in the inkjet printer. The paint is sprayed out of a positively charged nozzle. As all the paint droplets from the nozzle are positively charged, they repel each other and spread out into a paint mist. The car or object to be painted is connected to the negative so that the charged paint is attracted to it. The paint hits the car evenly in all places, including the hidden parts and edges. This gives an even coat all over.

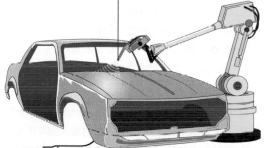

positively charged paint nozzle

car body connected to negative

How would the quality of paint finish be affected if:

d) The voltage between paint and car was increased?

e) The car was also positively charged?

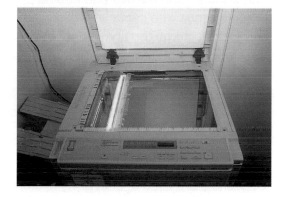

The photocopier

Photocopiers use a bright lamp to 'scan' a page. The light is focused through the page to be copied. The paper absorbs some of the light but only where there is text or images on the page. The rest of the light beam goes through the page and onto a drum plate. The plate has been charged up. Any light hitting the plate knocks the charge off (makes it 'leak' away). This process leaves a charged image of the page on the plate. The drum is then passed through negatively charged ink dust. The ink sticks to the charged image and is then pasted onto a piece of plain paper. The page is heated to make the ink stick properly. The finished copy is now complete, all in less than a second! Laser printers work in a similar fashion to this.

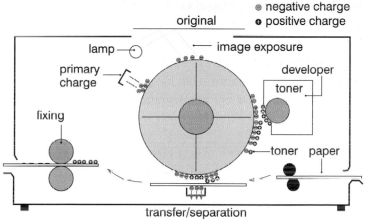

⊝ negative charge
⊕ positive charge

original

lamp

image exposure

primary charge

developer

toner

fixing

toner   paper

transfer/separation

f) Describe how the photocopier works in your own words.

Remind yourself!

1 Copy and complete:

Inkjet printers have a print ...... that charges the droplets of ...... as they leave. The ink droplets then pass between ...... and negative plates. The ...... deflect the ink droplets onto one exact part of the paper, forming part of the final image. ...... form a charge image of a page on a plate drum. The drum is then inked and the plate image ...... onto a piece of plain paper. The paper is ...... to produce the final copy.

2 Photocopiers use lenses and mirrors to help to focus the image of the page onto the plate. Why do you think they sometimes give blurred copies?

3 Most inkjet printers can give colour prints. How do they manage to print pictures up to photo quality using just three colours in the print cartridge?

You have seen the effects of electrostatic charge
suddenly discharging; **lightning**.
Have you ever had static electricity discharging through your hand?
Luckily most people are never hit by lightning
but we have all felt some form of static shock at one time or another.

Capacitors are devices often used in electrical equipment.
They store charge and then release it quickly when it's needed.
When you unplug computers or videos, be careful!
Never touch both the live and neutral pins at the same time.
When a circuit with capacitors is turned off, they discharge
any stored energy they have. Touching the plug pins will allow
the stored charge to discharge through you, giving you a small shock.

capacitors

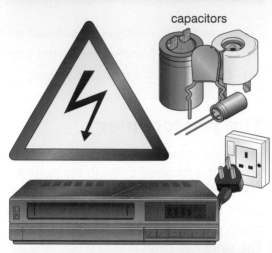

a) Have you ever felt a static shock? Where?

Stun guns

Capacitors are used in stun guns to store charge.
When the stun gun is used to resist attack, charge from
the capacitors discharges. The energy is released
in the form of a very high voltage 'pulsing' shock.
This shock has an effect on the nervous system.
It can leave a person stunned for a few minutes.
The police in the UK are investigating the use of stun guns
as a possible restraining weapon.

b) What are your opinions on the use of stun guns by the police?
Discuss your ideas with a friend.

Lightning conductors

Clouds can become negatively charged. This affects the surface
of the Earth beneath the cloud. It becomes positively charged.
When the potential difference between the cloud and the ground
is large enough, some of the air molecules in between can become
ionised. (Positive and negative ions are formed.)
Suddenly a current flows (lightning) as the positive ions go
to the cloud and the negative ions and any electrons go to the ground.
Lightning conductors are points of metal placed on the highest part
of a building. Their main role is to attract the lightning so that the
current passes down their cables to Earth rather than through the building.
They can also help to discharge the cloud slowly, stopping any lightning.

lightning conductor

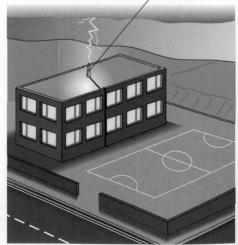

c) See how many lightning conductors you can spot on your way home.

The electrostatic precipitator

Factories and power stations have chimneys for getting rid of steam and waste smoke (tiny dust particles). The smoke can contain many pollutants that are bad for the environment. These need to be stopped from getting into the atmosphere.

The dust particles can be removed using an **electrostatic precipitator** placed in the chimney. This device has two large negatively charged plates opposite each other on the walls of the chimney. In the middle of the chimney there is a positive plate. See diagram opposite.

As smoke dust passes up the chimney, the positive plate makes the particles become positively charged. They are then attracted to the negative plates, which collect the dust particles, stopping them from escaping into the atmosphere.

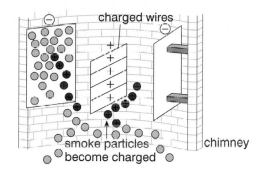

d) What happens to the dust that doesn't get charged up by the positive plate?

Charge and fuel

We all know that a spark near petrol is dangerous. The petrol engine works by mini explosions caused by sparks in petrol vapour.
You must have seen the signs in petrol stations banning the use of mobile phones! This is because a phone could possibly cause a spark near petrol vapour.
That could lead to a new petrol station being built earlier than the fuel company originally intended!

Petrol flowing through a pipe causes a build-up of static charge. To reduce the chances of a spark, car fuel tanks are always 'earthed' (connected to the body of the car). This allows any charge to flow away before enough static can collect to form sparks.

Remind yourself!

1 Copy and complete:

When static ...... builds up, it will always try to discharge. As charge discharges, sparks jump and you can get an electric s......
Lightning is a large ...... from clouds that are discharging to the Earth.
Connecting fuel tanks to vehicle bodies reduces charge build-up stopping ...... near ...... vapour.
Capacitors can give a shock as their stored charge discharges. Don't touch the live and neutral pins of a removed plug just in case.

2 Stun guns can kill people who have a weak heart or those fitted with a pace maker. Would this fact change the conclusion you reached in question **b**)?

3 Use an encyclopaedia or the Internet to do some research into the work of Michael Faraday.

Why does electricity flow through us when we are not made out of metal?
The human body conducts electricity because we have
charged particles on our skin and within our body.

Why must you never have a mains radio near a bath full of water?
When a radio falls into a bath, the water touches the internal
circuits including the live cables. Any charged particles within
the water will form an electric current through the water.
If you are in the bath, you feel the current flowing, but not for long!

What happens to a car battery when the acid is drained out?
No charge can flow inside the battery, so it becomes useless.

Ions as charge carriers

At the start of this chapter (page 81), you saw that electric charge
can flow through a liquid. This is because liquids can contain ions.

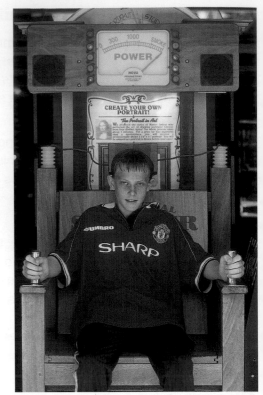

A fairground 'electric chair' game
(not a real one!).

> **a)** Can you remember what an ion is?

Ions are atoms or molecules that carry a charge.

> Negative ions have more electrons than protons. (More negative charge.)
> Positive ions have less electrons than protons. (More positive charge.)

> **b)** A liquid has a potential difference across it.
> Describe the movement of any positive and negative ions.

Liquids can contain no ions. This would mean that all the atoms
and molecules within the liquid are neutral. For example,
de-ionised water is used to top up car batteries.

Many compounds dissolved or melted, contain ions.
The extraction of aluminium from its ore involves
melting the ore. This is done in a cell that has a potential
difference between two electrodes. When you melt
the aluminium ore, positive aluminium ions become free
to move. They collect at the negative electrode and are
discharged as aluminium metal.

> **c)** Have you ever made copper from a blue solution (*often copper sulphate*)?
> **d)** Metals always form positive ions.
> Which contact will they collect at?

Electrolysis

The process of electrolysis happens when a liquid has
a potential difference across it. Any ions present flow
to the oppositely charged contact (electrode).
The liquid is broken down in the process.

Look at the diagram on the right.
There are two types of particles shown.
the positive ions (*red* ●)
and the negative ions (*blue* ●).

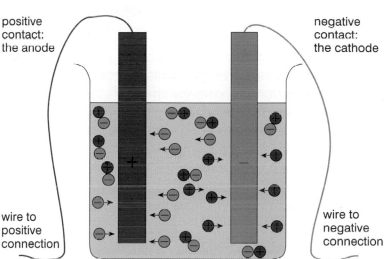

When the contacts are connected to
a potential difference, a charge flows.
The ions are attracted towards the electrodes.
The positive ions move to the negative electrode,
called the **cathode**. The negative ions move
towards the positive electrode, called the **anode**.
The neutral particles are not affected
by the potential difference between the electrodes.
They just continue to move about slowly within the liquid.

e) If the potential difference between the cathode and the anode
was increased, what would happen to the speed of the ions?

When ions arrive at an electrode, what happens next?

Cathodes give positive ions electrons until their charge is neutral again.
Anodes take electrons from negative ions until their charge is neutral again.

The atoms or molecules that have collected at the electrodes
become neutral.
They are always simpler substances than those in the main liquid.
If they are gases, they will bubble out of the liquid.

Remind yourself!

1 Copy and complete:

...... ions have less electrons than protons. They
move to the ...... electrode (called the ......)
during electrolysis.
Negative ions have more ...... than protons.
They move to the ...... electrode (the ......)
during electrolysis.
Metals always form ...... ions during electrolysis.

2 The process of electroplating can be used to
create a very thin layer of a valuable metal on the
surface of a cheaper metal.
Which electrode do you think the object to be
plated would become inside the plating liquid?

3 Water that contains no ions is called de-ionised
water.
Find out what the difference is between
de-ionised water and distilled water.
Which should you use to top up a car battery?

Summary

Static electricity is charge that can't flow away through a conductor.
It is stationary on the surface of an object.

The atom
Everything is made out of tiny particles called atoms. They contain:
- **Neutrons** that have no electric charge.
- **Protons** that are positively charged.
- **Electrons** that are negatively charged.

Protons and neutrons are both found inside the nucleus of an atom.
They are very difficult to remove from atoms.
Electrons orbit the nucleus of the atom.
Electrons can be dislodged from an atom to become free electrons.
Neutrally charged atoms have the same number of protons and electrons.

Charging and discharging static electricity
Electrons are the only charge carriers that can move in solids.
The surface that has lost electrons becomes positively charged.
The surface that has gained extra electrons becomes negatively charged.

When discharging an object to the ground:
If the object is positively charged, the electrons flow from the ground to it.
If the object is negatively charged, the electrons flow from it to the ground.

Static electricity has its uses
Inkjet printers use static electricity to focus tiny charged droplets of ink
onto the paper in exactly the right place to give clear characters.

Laser printers, photocopiers, stun guns, spray painting, lightning conductors
and electrostatic precipitators (for factory waste gases) all use static electricity.

Ions and electrolysis
Ions are either charged atoms or charged molecules.
(Molecules are particles formed when two or more atoms join together.)

A positively charged ion has more protons than electrons.
It will have lost some of its electrons, making it more positive.

A negatively charged ion has more electrons than protons.
It will have gained some electrons from another atom.

Electrolysis is a process where a potential difference is placed across a liquid.
Any ions in the liquid will be attracted to one of the electrodes (electrical contacts)
and repelled by the other.
The positive ions move to the negative electrode, called the **cathode**.
The negative ions move towards the positive electrode, called the **anode**.

Questions

1 Copy and complete:

  i) Static electricity is ...... charge.
  It is charge that can't ...... away.

  ii) Everything is made out of tiny particles called ...... They contain ......, neutrons and ...... Protons have a ...... charge. Electrons have a ...... charge and they can become dislodged from an atom.

  iii) A positively charged surface of an insulator has ...... electrons. A ...... charged surface has gained electrons.

  iv) Static electricity is used in a range of common devices including ...... printers, ..., spray painting and ...... guns.

  v) An ion is an atom (or molecule) with an unbalanced ...... A ...... ion has less electrons than ...... A negative ion has more ...... than protons.
  A ...... across a liquid that contains ions, will cause the ...... ions to move to the negative electrode (......). The negative ions will move to the positive ...... (anode) This process is called ......

2 Make sketches, highlighting the charges, for each of the following:

  i) A negatively charged rod being brought close to a neutrally charged rod.

  ii) A negatively charged rod touching a positively charged rod.

  iii) A balloon rubbed on a jumper and then stuck onto the wall.

  iv) The charges collected on a TV screen after being on for a few hours, affecting the hair on your arm.

3 Look back a few pages to the section on the uses of static electricity. Describe in your own words using a sketch diagram how:

  i) An inkjet printer works.

  ii) A photocopier works.

4 You might have noticed signs in petrol stations banning mobile phones or other electrical equipment. Any sparks near fuel vapour could cause an explosion.
The signs used are not always that 'eye catching' and some people ignore them.
Design your own banner that tells people how dangerous sparks near fuels can be!

5 Aeroplanes being refuelled are connected to their supply tanker.
Explain why this would prevent charge building up inside the planes' fuel tanks.

6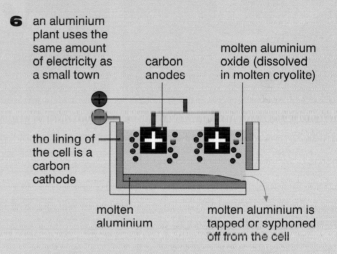

an aluminium plant uses the same amount of electricity as a small town

carbon anodes

molten aluminium oxide (dissolved in molten cryolite)

tho lining of the cell is a carbon cathode

molten aluminium

molten aluminium is tapped or syphoned off from the cell

Look at the diagram above. It represents an electrolysis plant used to extract aluminium metal from its ore.

  i) Which electrode is the aluminium metal forming at?

  ii) What type of ion, positive or negative do the atoms of aluminium form?

  iii) Oxygen ions travel to the anode (positive electrode). What type of ion, positive or negative, do atoms of oxygen form?

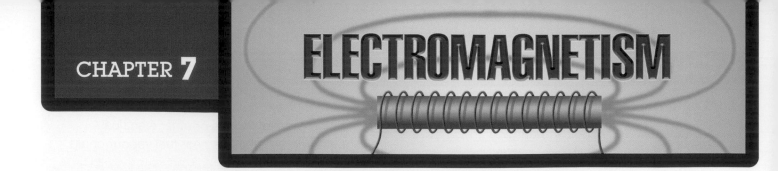

CHAPTER 7

ELECTROMAGNETISM

▶▶▶ 7a Magnetism and electricity

The alien Magnuts from the planet Magnex are on their way to Earth! They are after our magnetic elements.

To prevent humans from stopping them in their evil quest, they are going to bombard the whole of the Earth's surface with a 'Demag' beam.

It will stop everything that relies on magnetism to work. How can we defend ourselves?

Hopefully the Magnuts from planet Magnex will go and bother some other planet before us!

a) What are the three magnetic elements?

b) Why would stopping magnetism be so disastrous for us?

mini disc player

mobile phone

computer

washing machine

You have probably been learning about magnets since primary school.

You are most likely an expert at fridge magnet reorganisation!

You might even be able to reorganise the words below into an important expression:

like   attract   poles   repel   unlike   magnetic   poles

Electric current and magnetism are linked.

Electric motors, speakers, microphones, TVs and radios all work because of the **magnetic effect** of **electricity**.

Make a list of machines that:

c) Use electric motors   **d)** Use speakers   **e)** Use radio/TV waves

Magnets and magnetic fields

A magnet affects the space around it, turning it into a **magnetic field**. This is a region where the magnet affects any other magnets and magnetic materials that come close.
A diagram of a magnetic field shows the direction and shape of the field with arrows and **field lines**. The field lines are drawn close together to represent a strong field and spaced further apart for a weaker field. The field line arrows always point from north to south. A tiny compass placed in the field would line up with the arrows, pointing away from the north pole towards the south pole.

bar magnet field   field lines and arrows together show the strength and direction of magnetic field

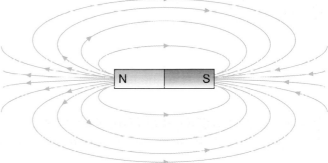

> **f)** Where is the magnetic field strongest in the diagram of a bar magnet shown above?

Permanent magnets are normally made out of steel. It is called a 'hard' magnetic material because once magnetised, it holds its magnetism well. Steel is not an element but it contains **iron** which is one of the three magnetic elements. The other magnetic elements are **nickel** and **cobalt**. Iron can be used for magnets but it loses its magnetism easily so it is a 'soft' magnetic material. Magnets attract or repel other magnets, depending upon which way their poles are facing.

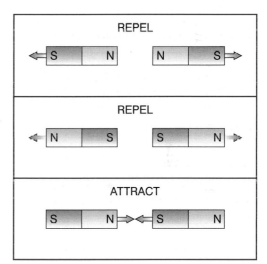

The electromagnetic effect

> A current flowing through a wire **induces** (makes) a magnetic field around it.

As the current flows through the wire, the space around it becomes a circular magnetic field. This is called the **electromagnetic effect** of a current. The field will be either clockwise or anticlockwise around the wire, depending on which way the current is flowing. *Use your **right** thumb!*
Hitching a lift, if your thumb points in the direction of the current, your fingers show the direction of the field.

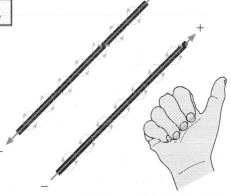

Remind yourself!

1 Copy and complete:

A ...... flowing through a wire induces (......) a circular magnetic field around it.
Line up your ...... thumb with the current (conventional; +ve to –ve), your ...... will trace the path of the field.
Sketches of magnetic fields have ...... that point from ...... to south. Tightly packed magnetic field lines represent a ...... field.

2 Compass needles are bar magnets. The tip of their arrow is a north pole.
Why do compasses always point to the north of the Earth if 'like poles repel'?

3 Write a poem or a story describing a day without electricity or magnetism.

How often do you:
- listen to the top 20 on your hi-fi speakers?
- sing *perfectly* through a microphone of a karaoke machine?
- ring a door bell or hear the telephone ring?

Speakers, microphones and bells all use **electromagnets**.

Turning a car's ignition key triggers a **relay** which is like a remote switch. The relay uses an **electromagnet** to work a starter motor which, hopefully, starts the engine.

When you use a power-pack in class, you are told which voltage to set it at. Have you ever noticed that power packs always seem to need the **circuit breaker** (the little button that 'pops' out) to be pushed back in again before they work? They must all be faulty, of course! It can't be anything to do with students trying to 'trip-out' the power packs by using too high a voltage! Circuit breakers use electromagnets.

circuit breaker

The solenoid

An electric current flowing through a wire produces a circular magnetic field around the wire. A wire twisted into a **coil** is called a **solenoid**. As a current flows through the solenoid, the magnetic field around each part of the wire adds together. The solenoid is an electromagnet. Its magnetic field is just like a bar magnet. To find the north and south poles of the solenoid, match up the letters N and S with the current direction.

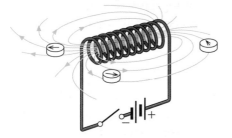

> **a)** What would happen to the poles of the solenoid coil in the picture if the direction of the current was reversed?

The strength of an electromagnet

You have probably constructed electromagnets during class practicals.

> **b)** Sketch a simple electromagnet and label it.
> **c)** What factors affect the strength of an electromagnet?

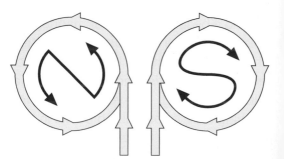

An iron **core** can be added to a solenoid electromagnet to make it stronger. The core would be inserted into the middle of the coil. It would help to concentrate the magnetic field making the electromagnet much stronger. Cores can be made out of any magnetic metal.

> Rewrite the mixed up statements below about electromagnets.
> **d)** stronger the current electromagnet through the coil Increasing makes the.
> **e)** increases Using a iron core the of an soft electromagnet strength.
> **f)** The number the of turns on the greater coil, the the electromagnet stronger.
> **g)** as cores Only metals work inside magnetic electromagnets.

Electromagnets in speakers

A speaker has a cone for pushing air. This makes sound waves.
The cone has a coil of wire attached at its small end.
A varying electric current flows through the coil,
making it an electromagnet. The coil is surrounded
by the magnetic field from a permanent magnet.
The two fields interact making a **motor force** which moves
the cone backwards and forwards.
(The electromagnetic motor force is described in more detail on page 97)

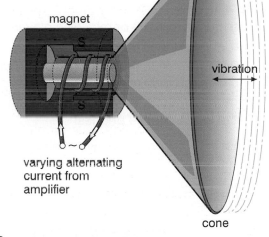

magnet

vibration

varying alternating
current from
amplifier

cone

h) To make the speaker's electromagnet move more,
which control on the hi-fi would you need to adjust?

Circuit breakers

Many appliances and household mains control units use circuit breakers.
They can immediately stop all electricity flowing if a fault develops.
A circuit breaker has a spring loaded contact and an electromagnet.
The current flowing in the circuit goes through both the contact
and the electromagnet. If for some reason too much current starts
to flow through the circuit, the electromagnet gets stronger.
It *pulls* the contacts apart stopping the current completely.

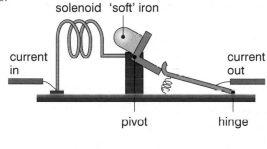

solenoid   'soft' iron

current
in

current
out

pivot   hinge

Relays

Relays have many uses as remote switches.
They are often used to control components in cars that require
a large current. Behind the dashboard of a car there is not much
space, so thick wires carrying high currents would not be much use.
The small switches on the dashboard control tiny electromagnets
inside relays elsewhere in the car. When magnetised,
the electromagnets attract a contact which completes the larger
current circuit.

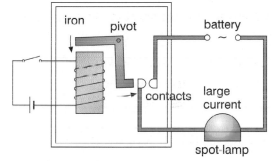

iron   pivot   battery

contacts   large
current

spot-lamp

This is how a small switch controls massive
spot-lamps!

Remind yourself!

1 Copy and complete:

The ...... field around a series of wire coils
(a s......) is the same shape as the field around
a ...... magnet. If the current direction is ......, the
magnetic poles of the solenoid are also reversed.
A solenoid is a simple ......
Factors that affect the strength of
electromagnets:
the ...... of turns of wire
the ...... through the wire or the supply voltage.
the type of ...... used inside the wire coils.

2 Speakers and microphones work using the same
principle.

i)   Draw energy flow diagrams for both a
speaker and a microphone.

ii) What do you think would happen if you
connected a speaker up to an amplifier
through the microphone plug and then spoke
into it?

3 How would you adjust the design of the circuit
breaker to allow it to stand a larger current?

Can you imagine a world without TV?
A life with:

- no music from CD or mini disc players
- no videos or DVDs to watch
- no microwave ovens for quick snacks
- no cars, motorbikes or buses for transport.

A place where not even computers and the Internet worked!
Worst of all, no mobile phones!

Without the relationship between electricity and magnetism, motors would not turn.
Dynamos and alternators from cars, motorbikes and buses would not generate electricity.
Speakers would be silent (not a bad thing sometimes). Microwaves and radio waves
would not exist. All the machines listed above would be useless!

The World would be very different
if magnetism were not linked to electricity!

a) Name five machines with motors that you could not live without.

Conductors and magnetic fields

The electromagnetic effect is when electricity flowing through
a conductor induces (makes) a magnetic field around it.
All motors rely on this effect to work.

So what happens when a current carrying
wire is near a magnetic field from a magnet?
The two magnetic fields interact.
In some places the two fields will add together
creating a much stronger field. In other places
the two fields will cancel.

b) In the diagram on the right, where is the
magnetic field adding together?

c) Is the magnetic field weakest or strongest at
point A in the diagram?

d) Imagine that the wire in the diagram
was free to move.
Would it be forced to move up or down?
Try to explain your answer.

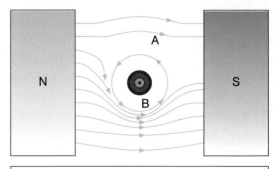

A wire drawn in cross-section.
Its current is flowing 'out of the page'.
Think of an arrowhead.

If the current is going in the opposite
direction, 'into the page', it is shown
with an arrow tail.

Fig. 1.

The motor effect

If your answer to question d) was that the wire would move upwards towards the weaker field, well done! The wire will move away from the stronger field towards the weaker field (upwards in this case).
This motion is called the **motor effect**.
Above the wire, the two fields are cancelling out as they are opposed (pointing in opposite directions).
Below the wire, the two fields are adding together forming a stronger magnetic field.

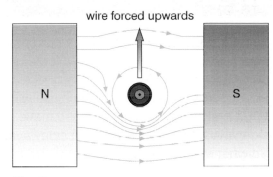

wire forced upwards

Fig. 2.

e) What would happen to the direction of the force on the wire if the magnet's magnetic field was reversed?

If magnetic poles were placed the other way around, the magnetic field would point in the opposite direction. The two fields would now be stronger above the wire and weaker below it. So the force on the wire would be downwards.

> For a current carrying conductor in a magnetic field: reversing the direction of the magnetic field, reverses the direction of the force on the wire.

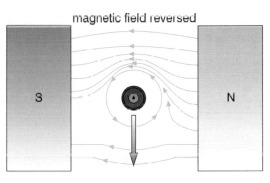

magnetic field reversed

Fig. 3.

f) If the direction of the current through a wire within a magnetic field was reversed, what do you think would happen?

In Fig. 4, the field is in the same direction as Fig. 1. The direction of the current has been reversed. The wire is being forced down now because the stronger magnetic field is above the wire and the weaker field below. (The same effect as Fig. 3.)

> For a current-carrying conductor in a magnetic field; reversing the direction of the wire's current reverses the direction of the force on it.

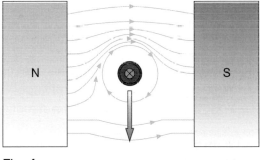

current direction reversed

Fig. 4.

Remind yourself!

1 Copy and complete:

The ...... field around a current-carrying conductor interacts with other ...... fields.
In some regions the fields will ...... up while in other regions they will subtract from each other.
The fields interacting produce a force, on the conductor, called the ...... effect.
If either the direction of the ...... or the magnetic field is ......, the force also changes direction.

2 Sketch a diagram similar to Fig. 3 but with the current going away from you (into the page). Label the direction of the force.
Which other Fig. has a force in the same direction?

3 If the size of the current was increased, do you think it would affect the size of the force?

In the last section you saw that a current-carrying wire near a magnetic field produces a force on the wire.
So what's the big deal!
How does this make all the machines with motors work?
What if a magnet can make a wire move up, down or even sideways, motors go round, not sideways!
So what makes a motor turn?

The **motor effect** is the force induced in a wire carrying a current within a magnetic field.
Increased current produces a larger magnetic field around a conductor. A larger magnetic field from a conductor interacts more with other magnetic fields producing a larger force.

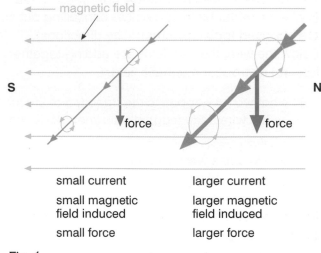

Fig. 1.

If the current is increased in the conductor, the motor effect force on it is increased.

a) How would changing the strength of the magnetic field surrounding a conductor affect the size of the force?

Fleming's left hand rule

What about the direction of the force and the motion induced?
Fleming's left hand rule is an easy way of working out the direction of the force (and the motion induced) on a current carrying conductor placed in a magnetic field.

Your **f**irst finger lines up in the direction of the magnetic **f**ield.
Your **c**entre finger (or se**c**ond finger) in the direction of the **c**urrent.
Your thu**m**b will show you the direction of **m**otion.

Line up your left hand for a current going up the page instead.
Keep the field in the same direction.

b) What happens to the direction of the force?

c) Is the new direction of the force what you would have expected?

d) Line up your right hand with the diagram. Does it work?

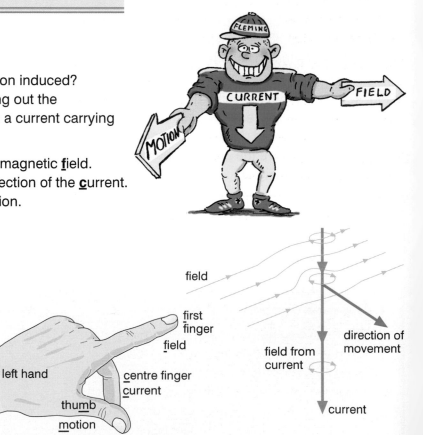

Fig. 2.

A simple electric motor

The most simple electric motor is based on a loop
of wire in a magnetic field. The loop or coil
is supported on an axle so that it can rotate.
Look at the diagram opposite:
As the current goes around the coil, section A–B
is given a motor effect force downwards.
C–D is given force upwards.

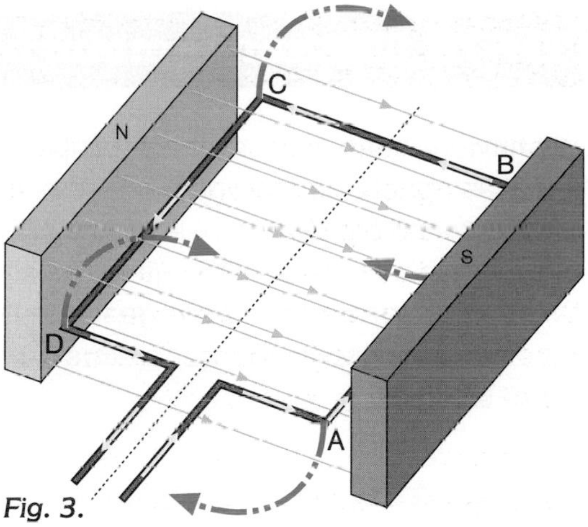

Fig. 3.

> **e)** Line up your left hand against the wire
> sections in the coil diagram.
> Do the directions shown for the forces on
> both section A–B and section C–D agree?
>
> **f)** Try to line up your left hand with sections
> B–C and D–A. What do you find?

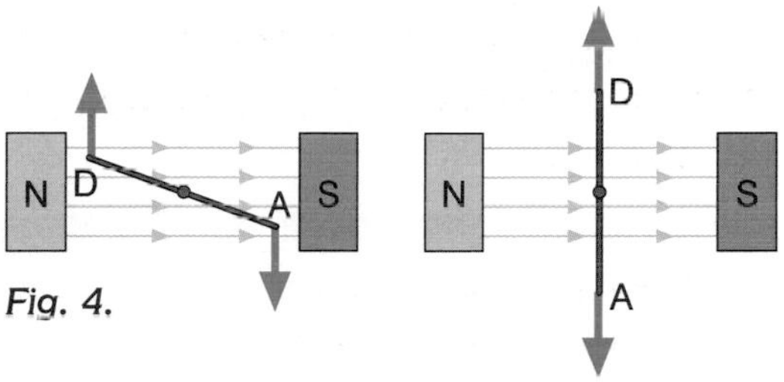

The sections B–C and D–A do not have any force acting
on them because they *line up with the magnetic field*.
They do not *cross the magnetic field lines*.
So what happens if the coil turns until it is upright?
As you can see in the diagram on the right,
the forces still act but they are now pulling
the coil apart instead of making it turn.
A–B would be stuck down at the bottom. C–D would
be stuck at the top. The coil stops turning.

Fig. 4.

Motors need to keep turning, not stop every half a turn!
A motor uses a special contact to the coil called a **slip ring**.
The slip ring changes the direction of the current through the coil
every half a turn of the motor.
As the motor coil spins, the contacts are continually
being made and broken at the slip ring.
This ensures that the direction of current around the coil
always produces a motor effect force in the same direction.
The coil keeps rotating in the same direction. Clever stuff!

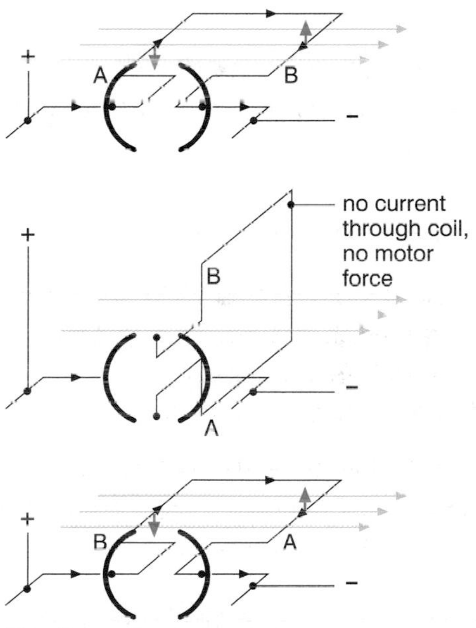

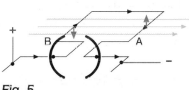

no current
through coil,
no motor
force

Fig. 5.

Remind yourself!

1 Copy and complete:

The size of ...... effect force on a conductor
depends upon both the amount of ...... flowing
through it and the strength of the ...... field the
conductor is in.

...... rings are used to ensure that the current
through a motor's coil always flows in the ......
direction.

2 Use Fleming's left hand rule to work out what
would happen to the coil in Fig. 3 if:

i) The current direction was reversed.

ii) The directions of both the current and the
magnetic field were reversed.

3 What do you think happens to a wire if it's moved
within a magnetic field.

We have millions of uses for electricity.
It makes motors turn. It lights houses. It's used for cooking.
In Chapters 1 and 2 we looked at the different uses and methods
of generating electricity. We generate electricity in power stations.
Ships, cars, lorries and buses generate their own electricity.
Turbines are used to turn generators but *how* does a generator
make electricity?

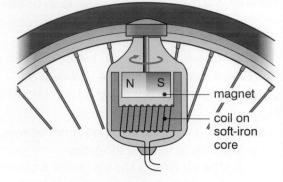

A dynamo on a bicycle.

magnet
coil on
soft-iron
core

Using motion to induce a voltage

The motor effect was discussed in the last section.
A current-carrying conductor placed across a magnetic field
will have a motor effect force acting on it.

a) Explain, with a sketch, how Fleming's left hand rule works.

The opposite is also true.

> A wire moved so that it crosses a magnetic field will have a voltage
> induced between its ends. This induces an electric current through it.

This relationship is called **electromagnetic induction**.
Electromagnetic induction occurs whenever a conductor
and a magnetic field are able to move relative to each other.
(One moves while the other doesn't.)
For electromagnetic induction to occur, lines of **magnetic flux**
must be cut. Magnetic flux is a way of describing the shape
and strength of a magnetic field as it points from north to south.
We represent magnetic flux with the magnetic field lines used in diagrams.
When a bar magnet is moved close to a solenoid coil, magnetic flux
is being cut by the wires of the coil. A current will be induced inside the solenoid.

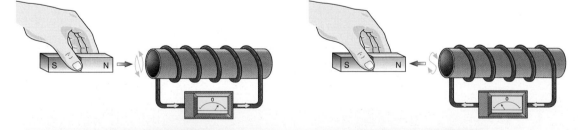

What do you think would happen to the direction of the induced current if:

b) The magnet in the diagram was being moved away from the solenoid?

c) The opposite pole of the magnet was moved towards the solenoid?

If the magnet was being moved away from the solenoid, the induced
current in the solenoid would be in the opposite direction.

> Reverse the direction of motion, the direction of the induced current also reverses.

The same would happen if the magnetic field was reversed by turning the magnet around.

Generating electricity

There is a rule which relates some of the fingers and the thumb
from the **right hand** to electromagnetic induction.

d) Who do you think worked out the rule for the direction of induced current?

e) Sketch a right hand and label the first finger, centre finger and thumb.

Electricity is generated using generators which are a bit like
huge motors working backwards.
A simple motor disconnected from a voltage supply and then
turned will generate electricity. The coil of wire inside it will
be cutting lines of magnetic flux so a current will be induced.

The diagram on the right is very
similar to Fig. 3 on page 99
Here the coil is being turned inside
the magnetic field.
An electric current is being induced
in the coil.

f) Line up your left hand against
the diagram. Do you agree with
the direction of motion?

g) What is the main difference between
this diagram and Fig. 3 on page 99?

Michael Faraday was the first to define the following points.

The size of the induced voltage depends on:
  the strength of the magnetic field
  the speed of the conductor or coil through the field
  the amount of conductor or number of turns of wire on a coil within the field.
(The greater the induced voltage, the larger the current.)

Remind yourself!

1 Copy and complete:

When either a ...... or a magnetic field move near
each other, a ...... is induced in the conductor.
This is because lines of magnetic flux are being
...... by the conductor.
This is called electromagnetic ......
The greater the induced voltage across the
conductor, the more ...... will flow along it.
The faster the lines of ...... are cut, the greater
the induced voltage.
Increasing the ...... of the magnetic field or the
amount of conductor/...... of coils inside the field
also increases the induced voltage.

2 Put your right and left hands into the positions for
Fleming's rules.
Bring them close together with your thumbs
upward.

  i) What is the only difference in the way the
     fingers/thumbs point?

  ii) Invent a way of remembering which hand is
      for induced motion and which is for induced
      current.

3 If you turn a simple dynamo in either direction, it
will light a bulb. Explain why the dynamo can
produce electricity working backwards.

Your portable CD player will probably have four 1.5 V batteries.
This adds up to a total supply voltage of 6 V.
So why can you also connect it to the 230 V mains supply?
The National Grid carries very high voltage electricity,
often more than 132 000 V.
So how can the supply to our homes be so much lower?
You are always having to connect your mobile phone up to its charger.
Why does the charger have a large black box at the plug?

All of these depend upon Transformers to change voltages.

Transformers

Transformers are used to reduce or increase voltages.
The have an input side (**primary coil**) and an output side (**secondary coil**).

A **step-up** transformer **increases voltage**.

Its output voltage is greater than its input voltage, but the current is reduced.

A **step-down** transformer **decreases voltage**.

Its output voltage is lower than its input voltage, but the current is increased.

a) List five common uses for step down transformers.
(Hint: the mobile phone charger uses a step-down transformer.)

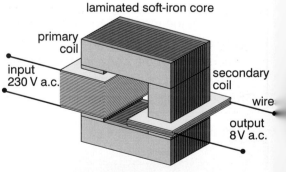

laminated soft-iron core

primary coil

input 230 V a.c.

secondary coil

wire

output 8 V a.c.

primary has 300 turns
secondary has 10 turns

stepdown ratio = 30:1
typical mobile phone charger,
230 V input, approx 8 V output

Transformers have two coils, the primary and the secondary.
The coils are placed very close to each other but are not connected.
Normally a laminated (thin layers stuck together) soft iron core
surrounds both coils. It helps to concentrate any magnetic flux.

Transformers rely on the principle of **electromagnetic induction**.
They do not have any moving parts. Instead, a continually
changing magnetic field around the primary coil induces
a voltage across the secondary coil. The changing magnetic field around
the primary coil comes from its **alternating current (a.c.)** supply.

Every time the alternating current in the primary coil changes direction,
the magnetic field around it reverses. This changing magnetic field
causes magnetic flux to cut across the wires of the secondary coil.
This induces a voltage across the secondary coil.

Alternating current (a.c.)
reverses its direction continually.

Direct current (d.c.)
flows from
positive to negative.
(*Conventional current.*)

b) What type of current is induced in the secondary coil?
c) What direction does *real* current flow in?

Transformers and the National Grid

You must have seen lines of pylons carrying high voltage electricity. They are most common in the countryside. In towns it is more common for electricity cables to be placed underground. The electricity produced at power stations has to be distributed to all our towns and cities by the **National Grid**. It is a network of pylons that carry very high voltage electricity cables. Power stations produce electricity at lower voltages than the voltage used in the National Grid.

d) What type of transformers are used to connect the power stations to the National Grid?

The electricity generated in a power station is put through a step-up transformer. This produces very high voltage electricity but at a much lower current. A lower current produces far less wasted heat energy as it passes through the cables of the National Grid.

e) What type of transformers are used at the town end of the National Grid?

The voltages used in large factories are still quite high (33 000 V), in offices and schools the voltages are often as low as 480 V. Domestic mains electricity is only 230 V (it used to be 240 V). Step-down transformers are used at each step to reduce the voltage down to the level required. Somewhere near to where you live there will be an electricity substation. At least one pylon line is attached to the station which will contain a number of step-down transformers.

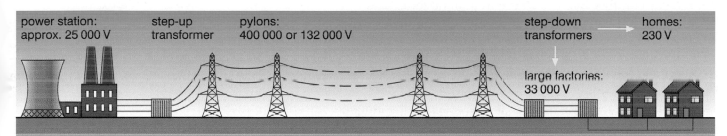

power station: approx. 25 000 V | step-up transformer | pylons: 400 000 or 132 000 V | step-down transformers → homes: 230 V | large factories: 33 000 V

Remind yourself!

1 Copy and complete:

...... are used to change voltages.
A ...... transformer increases the voltage and gives a lower current.
A step-down transformer ...... the voltage and gives a ...... current.
The ...... Grid transfers high voltage electricity.
The voltages are ...... up at power stations and stepped-...... at the other end to supply cities and towns.

2 Electric welding machines are really just big transformers.
Do you think they step-up or step-down the mains supply voltage?

3 How can a bird stand on an electricity power line if they carry very high voltage electricity?

4 Electricity substations are very dangerous places. Design a warning poster to stop children from entering them.

Summary

Magnets affect the space around them, turning it into a **magnetic field**.
A magnetic field diagram shows the direction and shape of the field
using arrows and **field lines**.
Field lines drawn close together represent a strong magnetic field.
The field line arrows always point from **north** to **south**.

An electric current flowing through a wire **induces** (makes) a circular shaped
magnetic field around it. This is called the **electromagnetic effect** of a current.

Electromagnets work because of the electromagnetic effect.
Speakers, microphones, relays and bells all use electromagnets.
A solenoid is an electromagnet in the shape of a coil.
Its magnetic field is just like that of a bar magnet.
To find the north and south poles of the solenoid,
match up the letters N and S with the current direction.

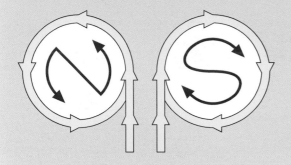

The **motor effect force** acts on a current-carrying conductor within a magnetic field.
Reversing the direction of the magnetic field, or the direction
of the current, reverses the direction that the force acts.
If the current is increased in the conductor, the motor effect force increases.

Electromagnetic induction is an induced voltage across a conductor.
The greater the induced voltage, the larger the current that can flow through the conductor.
Electromagnetic induction occurs whenever a conductor crosses a magnetic field.
Reverse the direction of motion, the direction of the induced current also reverses.
The size of the induced voltage depends on:
    the strength of the magnetic field
    the speed of the conductor or coil through the field
    the amount of conductor or number of turns on a coil within a magnetic field.

Transformers increase or decrease voltage.
A **step-up** transformer **increases** voltage. Its output voltage is greater than its input
voltage, but the current is reduced.
A **step-down** transformer **decreases** voltage.

The electricity generated in power stations is put through a step-up transformer.
This produces very high voltage electricity but at a much lower current.
A lower current produces far less wasted heat energy as it passes through
the cables of the National Grid.

Questions

1 Copy and complete:

i) Magnets create a ...... field.
Field ...... with arrows pointing from ...... to south are used to show magnetic fields.

ii) Electric ...... flowing through a wire produces a ...... magnetic field around the wire. This is called the ...... effect.

iii) ...... use the electromagnetic effect. They are used in many devices including ......, ..., ... and bells.

iv) When a current carrying ...... crosses a ...... field, a force acts on it.
The force is called the ...... effect.
Reversing either the ...... or magnetic ...... directions reverses the direction of the motor effect ......

v) When a conductor is moved so that it crosses a magnetic field, a voltage is ...... (made) across it. A current will flow through the conductor.
The induced ...... will be larger if the speed of the ...... is increased. Making the strength of the magnetic field or the amount of conductor in the field greater will also increase the voltage.

vi) Transformers increase or ...... voltages. A ......-up transformer increases voltage, a step-down ...... decreases voltage.

2 Describe any forces that exist between the following sets of parallel wires.

i) Two wires carrying the same current in the same direction.

ii) Two wires carrying the same current but one in the reverse direction.

iii) Two wires but one wire has no current flowing through it.
Use the principle of like and unlike magnetic poles plus the electromagnetic effect to help you answer.

3 Look back at the section on electromagnets.

i) Explain in your own words with a clear diagram how a circuit breaker works.

ii) Draw a sketch to show how a sound wave can make electricity through a microphone. (Microphones are very similar to speakers.)

4 Check that you remember Fleming's rules.

i) Think of a way of remembering which hand is used for induced motion and which is used for induced current.

ii) Try to find another way of remembering what the fingers and thumbs represent.

5 Describe the effect of the following changes on a simple electric motor.
(The motor is already rotating slowly clockwise.)

i) Increasing the current flowing through the coil.

ii) Decreasing the strength of the magnetic field surrounding the coil.

iii) Reversing the direction of the current flowing through the coil.

iv) Reversing the direction of the current through the coil and also the direction of the magnetic field.

6 Decide what type of transformer would be used for each of the following:

i) National Grid to local network.

ii) Mains electricity to arc welder.

iii) Mains to mobile phone charger.

iv) Power station to National Grid.

CHAPTER 8

DOMESTIC ELECTRICITY AND SAFETY

▶▶▶ 8a Three-pin plugs

Have you ever been abroad on holiday?
Just imagine this:

> You have just got to your hotel and you're in a rush for dinner.
> You need to use a portable iron on your clothes and you must
> charge up your mobile phone when you discover ...
> **They have the wrong type of plug sockets!**
> (Of course, you could always have taken an adaptor.)

Why do other countries not use the same type of plugs as the UK?
Some countries use two-pin plugs. Other countries use plugs with
three round pins or two flat and one round.

> **a)** Do you think the shape of the plug or the pins affects how well
> the appliance works?

Mains electricity systems use the same principles in all countries.
Supply voltages do vary slightly, in the UK we use 230 V.
240 V was the standard but this has been adjusted to fit in with the EEC.
The shape and type of plug is decided by each country but it has no effect
on how well the appliance works.

Fitting plugs

Have you ever had to wire up a three-pin plug at home?
When you buy a new electrical appliance it will have the plug already
fitted, ready for use. Until a few years ago, very few new appliances
had the plugs fitted. Normally you had to buy a plug separately and fit it yourself.

> **b)** Which statement below is more likely to be true?
> i) Electrical appliances have plugs fitted at the factory because it makes
> life easier for the customer.
> ii) Plugs are fitted at the factory to try to reduce the number of fires and
> accidents caused by poorly fitted plugs.
>
> **c)** Look at the four diagrams of plugs. Write down any faults that you spot.

Fitting a three-pin plug

When you are going to fit a plug, the first thing to notice is the type of cable fitted to the appliance. Lower current appliances like table lamps, radios and TVs often have a twin core cable. Appliances like vacuum cleaners, toasters and washing machines always have a three core cable.
The wires inside the cable are coloured.
They must always be fitted to the correct terminals of the plug.

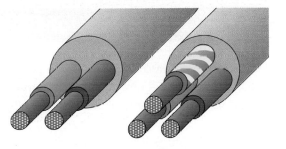

Colour of wire	Name of wire	Label under plug	Type of cable wire found in	Terminal it must be connected to (looking down at opened plug)
brown	**live** wire	L	Two and three core	Right hand terminal connected to fuse
blue	**neutral** wire	N	Two and three core	Left hand terminal
green/yellow striped	**earth** wire	E	Three core only	Top terminal. Longer one

Explain why each plug/cable part is made out of the material stated.
- **d)** Copper inner wires
- **e)** Plastic cable covers
- **f)** Brass plug pins and screws
- **g)** Plastic plug bodies

neutral wire

wires stripped carefully and connected so that all copper strands are inside contact tightly attached

earth wire, connected to longest pin

fuse, connected to live wire

live wire

inner cables cut to the right length so that they are not stretched or too loose inside plug

Remind yourself!

1 Copy and complete:

Mains electricity in the UK is supplied at ...... V. Inside a three-pin plug, the ...... wire connected to the fuse is called the ...... wire. The ...... wire is called the ...... wire. The ...... and yellow ...... wire connected to the longest pin is called the ...... wire.

2 Brass is a harder wearing metal than copper. Why would you not use copper for the pins on a plug?

3 Some older type small table lamps only have two wires and they are not coloured. Would you be able to wire them up to a modern three-pin plug?

Have you ever fitted a new fuse in a plug?
If so, did you use a new fuse of the same value as the old one?
Or did you just use any old fuse in the hope that it would work?

a) Why do plugs have a fuse?

There could well be some appliances in your house fitted with a 13 A
fuse that should have a smaller fuse. Any one of those appliances could
now be an accident waiting to happen!

b) List five appliances that you think should not be fitted with a 13 A fuse.

Fuses can be thought of as the 'weakest link' in an electrical circuit.
They are designed to *blow* if the current becomes too high.
When a fuse blows, the thin wire inside it melts.
This breaks the circuit so no more current can pass.
The fuse is the first part of the circuit to burn out, stopping
any of the circuits within the appliance from getting damaged.

c) Imagine that an appliance is fitted with a fuse that is too large.
What would happen if there was a fault and a large surge of current flowed?

When a larger fuse is fitted, it is not the first part of the circuit to blow.
A fault with the appliance causing a current surge might never occur.
It is very common to find plugs with too large a fuse fitted and often
nothing ever happens.
Just remember, with too large a fuse fitted, the current only has to surge once!

Example 1
A typical table lamp would need a 3 A fuse.
The flex for the lamp would normally be made out of thin, 5 A cable.
If a short circuit on the bulb holder occurred, the fuse would be tripped
as the current rises above 3 A.
If a 13 A fuse had been fitted, it would not be the fuse that would blow!
The whole flex would all act like a fuse and burn out!
Flexes are normally covered in plastic, plastic burns well!
Flexes are often near furniture and lie on carpets! **FIRE! HELP!**

Finding the right fuse

The fuse used in a plug should be a slightly higher value than an appliance's maximum operating current.
Household plug fuses are designed to blow at currents just above 3 A, 5 A or 13 A. For example, the correct fuse for a 4.5 A appliance would be a 5 A fuse, while a 5.5 A appliance would have to use a 13 A fuse.
Often appliances have a plate stamped somewhere with their current, voltage and power ratings. This would tell you the correct fuse to use.
The other way to find the correct fuse for an electrical appliance involves using the **electrical power equation**. (Equation number 9, see page 67.)

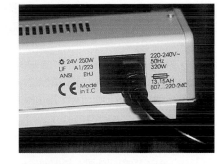

Appliance rating information.

	power = voltage × current		
Units	watts	volts	amps
	(W)	(V)	(A)

Example 2

The correct fuse for a mains 60 W table light;

  power = 60 W
  voltage = 230 V

Using a form of equation 9

$$current = \frac{power}{voltage} \div \text{ so current} = \frac{60\ W}{230\ V} \div = 0.26\ A$$

For a current of 0.26 A you could use a 1 A fuse.
The standard fuse sizes are 3 A, 5 A and 13 A
So the fuse you would use for a 60 W table light would be a 3 A fuse.

d) Copy and complete the table below to find the correct fuse sizes.
   You can get your answers by using the same equation as shown in Example 2.

Appliance name	Power rating	Supply voltage	Current used	3, 5 or 13 A fuse
TV set	460 W	230 V		
Computer system	690 W	230 V		
Vacuum cleaner	920 W	230 V		
Twin element fire	2760 W	230 V		

Remind yourself!

1 Copy and complete:

...... blow if the current through them gets too high. This ...... the appliance.
Household 3-pin plugs use either ... A, 5 A or ... A fuses. Always use the correct ......!
If too large a fuse is used on a low current appliance, it will not ...... A fault giving a current surge could cause a ......
Many appliances have the required ...... or their ...... rating stamped on a plate.

2 Fuses very rarely blow unless a fault causing a current surge has happened to the appliance.
What should you do if an appliance blows a new fuse soon after it is fitted?

3 Circuit breakers have many uses including stopping the mains supply. (Their operation was outlined on page 95.)
Find out about earth leakage circuit breakers and what they do.

Have you ever had a mains electricity shock?
If you ever have, you'll certainly remember it!
Even a minor shock can really hurt you.
Small shocks are sometimes fatal to people with heart conditions,
so it's best not to take any chances!

Every year people are killed by accidents involving electric shocks.
Those who do not die from electrocution can have severe burns
and other complications.

a) Which would you do if you saw someone getting an electric shock?
　i) Grab them away from the appliance that was giving them the shock.
　ii) Take the appliance away from them.
　iii) Turn off the appliance at the mains switch.
　iv) Leave them alone and rush off to get help as quickly as possible.

WARNING

MAINS ELECTRICITY DOES KILL.
RESPECT IT!

The most important thing to do when somebody is having an electric shock
is to stop the shock as quickly as possible, but:
- **Never touch them**. You will also get a shock.
- **Never touch the appliance**. Its fault will also give you a shock.
- **Always turn off the appliance at the mains plug SWITCH!**
  Even pulling the plug out can be risky,
  so if there is a switch, use it first.

Double insulation

Whenever you get an electric shock, the electric charge passing
through you is trying to get to the Earth. Electrical appliances
are designed so that the chances of you getting a shock are low.
There are two ways of protecting people from shocks:
double insulation and **full earthing**.
Double insulation is used on many lower current appliances
such as radios, TVs and hair driers. Electrical cables always
have an insulating covering of rubber or plastic.
A double insulated appliance will be made out of a material
that is an electrical insulator. This means that the user
of the appliance is protected from shock by both the cable
covering and the appliance's body.
Double insulated appliances do not have an earth
wire inside their cables.
They still have 3 pin plugs. We need the longer
Earth pin to open up the socket for the shorter
Live and Neutral pins.
This explains why you sometimes see double
insulated appliances with a plastic Earth pin.

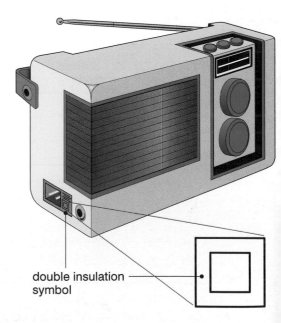

double insulation
symbol

Earthing

Full earthing is used on all higher current appliances as well as those with metal bodies. These appliances all have a three wire cable. The earth wire is connected between the plug and the case of the appliance.

Within a house's mains wiring, the other end of the earth wire is connected directly to the ground, the **Earth!** This is normally done through a stake in the ground. All metal plumbing, including heating pipes, are also connected to the earth circuit of a house. The earth wires provide a low resistance route for electric charge to get to the ground.

Pipes must be earthed.

b) What do you think would happen if a short circuit to the case of a washing machine happened just as you touched it?

If a live wire was able to work loose and touch the metal case of an appliance, there would be a short circuit. If the appliance was not earthed its case would remain *'live' and dangerous*. Touch the appliance and you would receive a bad electric shock as the charge flowed through you to the ground.

Earthed appliances enable charge from a short circuited case to flow directly to Earth. The resulting current would be large as there would be little resistance in the circuit. This would cause the fuse to be blown immediately or the circuit breaker to be tripped out.

Being safe with electricity

c) Look carefully at the picture on the right. In a table, list all the electrical hazards and why they are dangerous.

Remind yourself!

1 Copy and complete:

Electricity is very d......
Domestic electricity from the ...... carries easily enough energy to kill people.
If you come across somebody having an electric ......, always turn off the plug based ......
Do not ...... them or the appliance.
...... insulation and full ...... are safety measures used to protect people from ...... shocks.

2 Why is the earth pin on a three-pin plug always longer than the pins on the live or neutral contacts.
(Hint: think about safety first.)

3 Research and then design an information poster or a mini booklet.
It needs to tell people more about the hazards of carelessness with mains electricity.

If alternating current is always changing direction:
Why don't lights flash on and off as the current reverses?
When do the charges get to the other end of the circuit?

If you have had an electric shock from the mains, you might
remember that it seemed to pulse as it was hurting you.
With a mains shock you can actually feel the alternating current
changing direction as pulsing waves through you during the shock.

In Chapter 7, alternating current (a.c.) was introduced to explain how
a transformer works. The difference between a.c. and direct current (d.c.)
is that the current in a.c. reverses direction continually. Positive
to negative, then negative to positive, then positive to negative, etc.
In *conventional* d.c., the current only flows from positive to negative.

a) Do you know how often the current changes direction with a.c.?
(Hint: look at the data stamp on a piece of equipment.)

b) Electrons are the charge carriers through a conductor in a d.c. circuit.
What are the charge carriers in an a.c. circuit?

Frequency and a.c.

The word **frequency** is used to define how often something happens
each second. Normally referred to as the number of **cycles per second**.
Frequency is measured in Hertz (Hz). See page 170.
The number of times the current changes direction in the UK mains
is 50 times per second. So:

> The frequency of the mains supply in the UK is 50 Hz.

This means that in every second, the voltage between the 'live'
and the 'neutral' wire changes 50 times. The live wires potential changes
from 230 V above the neutral wires to 230 V below. Then it changes back again.
The neutral wire is really an earth connection but it is earthed at the power station.

Showing a.c. and d.c. on an oscilloscope

We use the **cathode ray oscilloscope** (normally called the **CRO**) to give a trace to show changes in voltage. It is like a mini TV screen. You might have seen a CRO being used to show a sound wave.

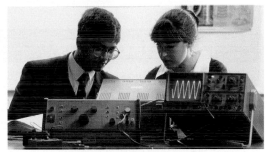

The CRO traces below are all for d.c. supplies

Trace 1

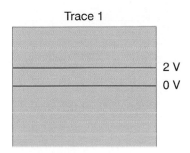

2 V
0 V

The output from a cell providing a d.c. potential of 2 V

Trace 2

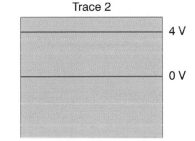

4 V

0 V

The output from a battery of two cells providing a d.c. potential of 4 V

Trace 3

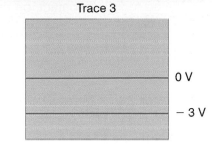

0 V

– 3 V

The output from a cell providing a d.c. potential of −3 V

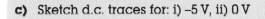

c) Sketch d.c. traces for: i) –5 V, ii) 0 V

The CRO traces below are all for a.c. supplies

Trace 1

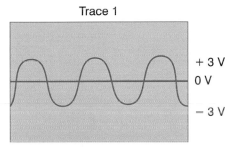

+ 3 V
0 V
– 3 V

The output from a 3 V a.c. supply

Trace 2

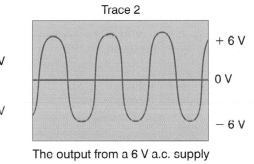

+ 6 V

0 V

– 6 V

The output from a 6 V a.c. supply

Trace 3

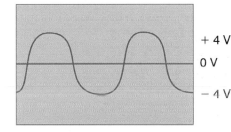

+ 4 V
0 V
– 4 V

The output from a 4 V a.c. supply at a lower frequency than Trace 1

Trace 3 has the lowest frequency, it has less cycles visible.
(*The CRO settings have not been changed*.)

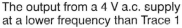

d) Sketch an a.c. trace for a 2 V supply at the same frequency as Trace 3 above.

Remind yourself!

1 Copy and complete:

Alternating ...... (a.c.) reverses its direction backwards and forwards continually.
The UK mains supply has a frequency of...... Hz. This means that the current reverses direction 50 times every ......
On a CRO, a.c. forms a transverse wave trace (up and down) while d.c. forms a ...... line trace.

2 It is possible to **rectify** a.c.
This is where every negative voltage drop is reversed so that only positive voltage changes occur. The CRO trace of rectified current is not smooth like true d.c.

i) What trace would you expect to see?

ii) Why would a mains supply mobile phone charger need a rectifier as well as a transformer?

Summary

In the UK, electrical connections use a three-pin plug.

The wires inside a plug must always be fitted to the correct terminals:

Live wire:

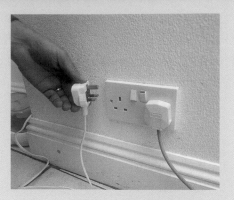

Colour code: brown   Plug label letter: **L**

Fitted to fused contact, on the right when looking into an open plug.

Neutral wire:

Colour code: blue   Plug label letter: **N**

Fitted to the contact on the left (opened plug).

Earth wire:

Colour code: green and yellow striped   Plug label letter: **E**

Fitted to the top (longer) contact (opened plug).

(Not all appliance cables have an earth wire.)

neutral wire —

wires stripped carefully and connected so that all copper strands are inside contact tightly attached

earth wire, connected to longest pin

fuse, connected to live wire

inner cables cut to the right length so that they are not stretched or too loose inside plug

Fuses are designed to *blow* if the current becomes too high.

They are intended to burn out first, preventing damage to an appliance.

Electrical power equation (Equation 9) is used to find the value of fuse required for an appliance.

	power =	voltage $\times$	current
Units	watts	volts	amp
	(W)	(V)	(A)

Electric shock. If you come across somebody having an electric shock:

Never touch them. You will also get a shock.

Never touch the appliance. Its fault will also give you a shock.

Always turn off the appliance at the mains plug SWITCH!

Even pulling the plug out can be risky, so if there is a switch, use it first.

Double insulation symbol.

A double insulated appliance will be made out of a material that is an electrical insulator. Users of the appliance are protected from shock by both the cable covering and the appliance's body. All the wiring within a house is connected to the ground by an **earth**. All metal plumbing is also connected to the earth circuit of a house. The earth wires provide a low resistance route for electric charge to get to the ground.

Mains electricity for domestic systems is 230 V a.c. at a frequency of 50 Hz. Alternating current (a.c.) changes direction, positive to negative and then negative to positive and then back again. It changes 50 times every second (its frequency).

Questions

1 Copy and complete:

i)   Plugs used in the UK have ...... pins. The ...... wire is brown and connects to the fused ...... The neutral wire is ...... and the earth wire is ...... and ......

ii) Fuses are used in ...... to protect the appliance. It is very important that the correct value ...... is fitted in a plug.

iii) ...... shocks can kill! You should never ...... somebody who is having an electric shock, ... the electrical appliance first.

iv) Double ...... appliances are made out of p...... or other insulators. Both the cable and the body of the ...... protect the user from ...... shock.

v)   Household wiring is always ...... to the ground. This is called the ...... The chance of a ...... from an earthed appliance is reduced. If a fault develops, the ...... will be directed straight to ......, not through the user.

vi) Mains electricity is supplied at ...... V It is an a.c. (...... current) supply. The current reverses direction ...... per second, its frequency is 50 ......

2 Using a piece of card or plain paper, write out a simple flow diagram to explain clearly how to wire up a three-pin plug. Include diagrams to help if possible. Make sure that the chart would work for a student that had never tried to wire up a plug before.

3 Many countries in the world only have two pins on their plug sockets.

i)   Which of the three contacts do you think they don't bother with?

ii) Do you think their system is still safe?

4 Work out the correct fuse (3 A, 5 A, or 13 A) for the following appliances:
Use the electrical power equation in the form, Power ÷ voltage = current.

i)   A 150 W spot-lamp, 230 V supply.

ii) An 800 W microwave oven, 230 V supply.

iii) A 1500 W kettle, 230 V supply.

iv) A 500 W video recorder, 230 V supply.

5 Write a story about a student who ignored the safety signs and broke into a local electricity substation.
What happened next?

6 Why do you think it is important for household plumbing to be earthed?

7 Look at the trace below from a cathode ray oscilloscope (CRO).

i)   Which trace represents a d.c. supply?

ii) Which trace represents the largest a.c. voltage?

iii) Which of the two a.c. traces shows the highest frequency supply (more waves per second)?

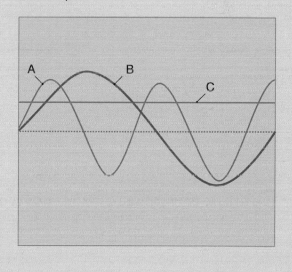

Further questions on Electricity

1 The circuit diagram shows a battery connected to five lamps. The currents through lamps **A** and **B** are shown.

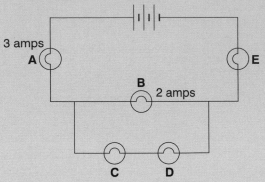

Write down the current flowing through

    (i)   lamp **C**.   (1)

    (ii) lamp **E**.   (1)

(WJEC 1999)

2 The graphs show the current through two devices when different voltages are applied across them.

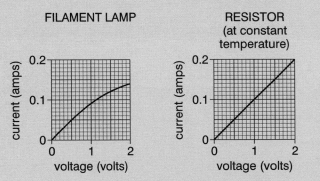

(a) Look at the symbols below and copy and label the ones for filament lamp and resistor.

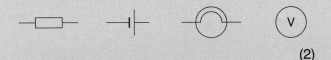

(2)

(b) Describe what happens to the current through the filament lamp and through the resistor as the voltage across them is increased.

    (i)   Filament lamp.

    (ii) Resistor.   (4)

(c) For very small voltages, the graph for the filament lamp is almost the same as for the resistor. Suggest a reason for this.   (1)

(AQA (NEAB) 1999)

3 Sally investigated the current through identical lamps (X and Y) in the electrical circuit below (Circuit 1). A1, A2 and A3 are ammeters.

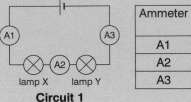

Ammeter	Ammeter reading (amperes)
A1	1
A2	
A3	1

Circuit 1

(a) Copy and complete the table above to show the reading on ammeter A2.   (1)

(b) Sally then rearranged the lamps and ammeters to make the circuit below (Circuit 2).

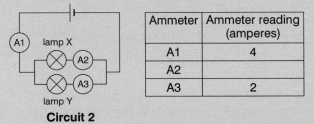

Ammeter	Ammeter reading (amperes)
A1	4
A2	
A3	2

Circuit 2

    (i)   Copy and complete the table above to show the reading on ammeter A2.   (1)

    (ii) She removes lamp X from its holder in Circuit 2.
        This leaves a gap in the circuit.
        Describe and explain what happens to lamp Y in this circuit.   (2)

(EDEXCEL 1998)

4   miniature circuit breaker   fuse
    earth leakage circuit breaker   earth wire

    Which of the above breaks the circuit when

    (i)   too large a current flows,   (1)

    (ii) a very small current flows through it?   (1)

(WJEC 1999)

5 (a) The diagram shows apparatus to demonstrate that copper sulphate solution conducts electricity.

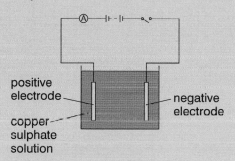

positive electrode

negative electrode

copper sulphate solution

Copy and complete each sentence by underlining the appropriate word.

In the wires, the electric current is the flow of **electrons/ions/protons**.

In the copper sulphate solution, the current is the movement of **electrons/ions/protons**.

Copper is deposited at the negative electrode by a process called **convection/electrolysis/precipitation**. (3)

(b) The diagram shows a device for reducing the amount of dust particles that are released from a factory chimney.

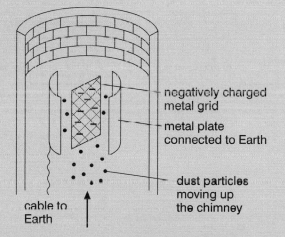

negatively charged metal grid

metal plate connected to Earth

dust particles moving up the chimney

cable to Earth

(i) The dust particles become negatively charged as they pass the metal grid. They then move towards the metal plates. Explain why the dust particles move towards the metal plates. (2)

(ii) When the dust particles touch the metal plates they lose their charge. Explain what happens to this charge. (2)

(AQA (NEAB) 2000)

6 The drawing shows a compass. The compass needle acts like a bar magnet and is free to rotate.

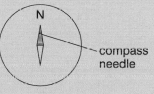

compass needle

(a) Wherever the compass is placed the compass needle will usually move round until the shaded part is pointing north. Explain why this happens. (1)

(b) A magnet is moved towards the compass as shown in the diagram.

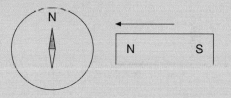

Describe the effect this would have on the compass needle. Explain why this would happen. (3)

(c) The diagram shows a coil attached to an ammeter.

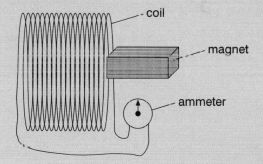

coil

magnet

ammeter

(i) When the magnet is moved into the coil the ammeter needle moves. Explain why this happens. (2)

(ii) Describe what happens when the magnet is moved out of the coil. Explain why this happens. (2)

(d) In a generator, electricity is produced when coils of wire are rotated in a magnetic field. Explain why this produces electricity. (3)

(AQA (NEAB) 2000)

7 The diagram shows the inside of a 3-pin mains plug.

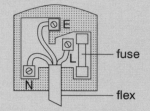

(a) Something is missing from the plug which makes it unsafe to use.
What is missing? (1)

(b) The table contains information about some electrical appliances.
Each one is connected to the 240 volt mains supply by a 3-pin plug.

Appliance	Power rating (W)	Current (A)	Fuse (A)
table lamp	60	0.25	1
lawn mower	750	3.1	5
electric drill	500	2.1	3
kettle	2500	10.4	

(i) What fuse should be used for the kettle? (1)

(ii) Explain how a fuse works. (1)

(iii) Why is it important to use the correct fuse for each appliance? (2)

(EDEXCEL 2000)

8 The diagram shows three positions of a coil as it spins through a magnetic field.

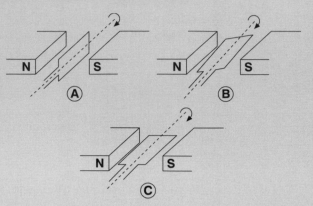

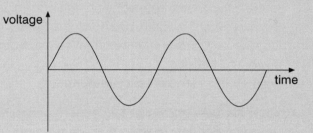

As the coil cuts through the magnetic field it produces the voltage shown on the graph below.

(a) State the position of the coil, **A**, **B** or **C**, when the voltage produced is

(i) a maximum, (1)

(ii) zero. (1)

(b) State **two** changes that could be made **to the coil**, to produce a bigger voltage. (2)

(WJEC 1999)

Section Three
Forces

In this section you will learn about the many different types
of force and how we use them within our world.
You will find out how to calculate the speed and acceleration
of moving objects and how to use friction to stop.
You will also study the formation of stars and think about the
existence of life beyond the Earth.

CHAPTER 9

Forces and their effects

▶▶▶ 9a Different types of force

You are an expert at using different types of force. Ever since you were born, you have been using forces to control everything about your life. Your first baby cries used a force to contract your vocal cords. Walking, talking, laughing and eating all use forces produced by your muscles. Managing to stand up without falling over depends on forces. Without the force of friction between the ground and your feet, you would fall over!

Boats float and planes fly only because of the way we have learnt to control and use forces. We have even managed to conquer the Earth's gravitational field, ensuring that we have satellite TV.

a) Give an example of how we use five of the forces below.

The ways that forces act

Gravity is a force of **attraction** between different masses.
Magnets can be either **attracted** or **repelled**, depending
on how their poles are arranged.

b) What do we call the forces of attraction and repulsion
between charged particles? Use the letters opposite.

E e t s t
t l i c
c a r o

With electrostatic forces, two positive charges will **repel** each other
while a positive charge will **attract** a negative charge.
Friction is a force between surfaces which acts to *resist* movement.
Friction occurs at the surface of different objects that are touching
such as a tyre on a road. Friction also acts on objects moving through fluids.
The **drag** on a swimmer or the **air resistance** on a table tennis ball
flying through the air are both examples of friction.

c) Explain, using ideas about friction, how a pencil marks a piece of paper.

A very Important polnt about the way forces act Is that:

> Forces between objects in contact are **equal** and **opposite**.

Put another way; every **action** *has an* **equal** *and* **opposite reaction**.
Whether this idea seems obvious to you or not, you know it!
You use it every moment of every day. The action of you sitting
makes your weight push down on the chair. The chair's reaction
pushes you back up with exactly the same force, balancing your weight.
If the force down was greater, you would sink into the chair.
If the force up was greater you would start moving upwards.
You don't because the forces are equal and opposite.
The same is true when you walk, run, hold a cup *or push a friend over!*

d) Sketch a basketball as it bounces, labelling the forces that act.

Remind yourself!

1 Copy and complete:

There are many different types of ......
Gravity is a force of ......
M...... and electrostatic are forces of both
attraction and ......
These forces can all act without physical ......
between the objects affected by the forces.
A contact-based force can only act if there is
direct ...... where the force is being applied.
Forces between objects in contact are always
...... and ...... When you jump, you push ......
on the ground. The ground pushes you back
with an ...... and opposite ......

2 Explain each of the following using the principle
that forces are equal and opposite.

i) A skater pushing against another moves
backward.

ii) Two firemen often hold the end of the hose.

3 Magnetism and electrostatics both involve forces
of attraction and repulsion. Gravity is only an
attractive force.
How do you think gravity acts on the surface of
other planets?

Have you ever fallen off a bike?
Balancing the forces on a bike so that you can keep upright
is a very complex process. Your brain has to continually adjust
how you lean to make sure that the bike remains upright.
Most children don't manage to balance on a bike until
they are at least five years old because it's so difficult.

Have you ever picked up a plastic cup that was much lighter
than you expected? The force you start to apply to the cup turns
out to be much too large and the drink starts to spill. Instantly, your
brain tells your arm to adjust the force so that the drink can be kept
in the cup.
We are experts at dealing with balanced and unbalanced forces.

Diagrams of forces

Forces are drawn in diagrams as arrows which show their direction.
Normally the length of the force arrow shows how large
the force is (but sometimes arrow thickness is used instead).
The **magnitude** (value or size) of the force is normally given as well.

> A box weighing 15 N is placed on a flat horizontal surface.
>
> **a)** Sketch the box and use force arrows to show its weight
> and its opposing upwards force from the surface.
>
> **b)** The box is now being pushed to the right with a 10 N force.
> The surface friction is 4 N. Add these forces to your sketch.

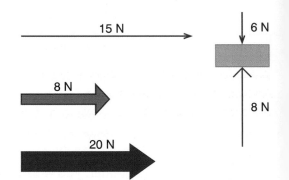

Ways of showing forces.

Newton's first law of motion

You might have heard about Sir Isaac Newton and the bump on his head
from a falling apple. Not only did Newton define the force of gravity,
he also gave us some laws to help explain forces and motion.
Newton's first law refers to **balanced** and **unbalanced** forces:

An object that is moving at a constant speed will keep moving.	*Balanced forces*
An object that is stationary will stay stationary,	*Balanced forces*
until an unbalanced force is applied to the object.	

Another way of putting it could be;
Nothing changes until you make it (apply a force).

> Use Newton's ideas to explain how:
>
> **c)** A pen held by a student often doesn't move for long periods of time.
>
> **d)** How a skateboarder can 'coast' on a flat surface without pushing much.

Balanced forces

A pair of balanced forces are equal but act in opposite directions,
so they cancel each other out.
Stationary objects have balanced forces acting on them.
They will not move until one force is reduced or an extra force is added.

Example, a Sea Harrier Jump Jet hovering

Its weight acts downward pulling it towards the ground.
The upthrust from its jets, pushing against the air below the jet, acts upwards.
The jet hovers above the ground as it has balanced forces acting vertically.

Think of an object moving at a constant speed.
One of the balanced forces is a **driving** force opposing any frictional forces.
The driving force from a car travelling at a constant speed cancels
out the forces from tyre friction and air resistance.

e) An astronaut on a space walk released by accident could demonstrate Newton's first
law of motion. What happens?

Unbalanced forces

When forces are unbalanced, a **resultant** force acts.
A resultant force is like the force left over after all the other forces
have been cancelled out. They make *things change*.

Unbalanced forces will:
- Make a stationary mass accelerate in the direction of the resultant force.
- Accelerate (increase the speed) of a mass if the resultant force
  is in the same direction as the mass's original motion.
- Decelerate (decrease the speed) of a mass if the resultant force
  is in opposite direction to the mass's motion.

The acceleration produced by a resultant force on an object depends
on the mass of the object. A larger mass will accelerate less than a smaller
mass given the same resultant force.

Remind yourself!

1 Copy and complete:

...... forces on a mass keep it ...... or continuing
to move at ...... speed and direction.
Un...... forces on a stationary mass will make it
move, accelerating in the ...... of the resultant
force. If the mass is already moving, unbalanced
...... on it will make it accelerate or decelerate,
depending upon which ...... the resultant force is
acting.

2 It is **impossible** for an object to have no forces
acting on it.
Do you think this statement is correct or wrong?
Explain your answer.

3 A helicopter can hover in much the same way as
the Jump Jet.
How can it hover and still spin around if balanced
forces are keeping it airborne?

Is mass the same as weight?
Do you really know the answer?
Most people spend their life believing that they are the same thing.
Other people say they know the difference but if you ask them their weight,
they will tell you they weigh $9\frac{1}{2}$ stones!
That's their mass, but what's their weight?
I bet you know your **mass** in stones.

a) Do you know your weight?

It's time to sort out this confusion!
Let's just imagine that mass = weight
So mass and weight are exactly the same thing, no difference.
Mass is matter, which depends upon the number of atoms present.
All things that exist have mass.

In deep space, well away from any planets or stars, there is little gravity.
This means that things are virtually **weightless** (they have **no weight**).
So you would be out in your space ship, floating about just like on TV, weightless.

But, if mass and weight are exactly the same thing, you would also be **mass-less**.
If mass = weight, then mass-less = weightless.
If you were mass-less, you would have no matter, no 'stuff' in you!

You would not exist!

You would go into space, then just disappear with your space ship, as if by magic!

Remember: mass is <u>not</u> the same as weight.

b) Write a short paragraph to remind you that mass and weight are not the same.

Mass

Everything is made out of atoms which contain neutrons, protons
and electrons (see page 80). All of these tiny particles have **mass**.
The mass of any object is the masses of all the atoms inside added together.

The amount of mass in an object is the same, wherever it is.

Whether the object is on the Earth, the Moon or in deep space, while the number
of atoms inside it doesn't change, its mass will always be the same.

All masses are attracted to all other masses.

There is a mass attraction force between your pen and the planet Jupiter!

c) Can you feel the mass attraction force between you
and the person sitting next to you?

d) What very strong mass attraction force is attracting you now?

Weight and gravity

> The mass attraction force between you and the Earth is called **gravity.**

The gravitational force of attraction between your pen and Jupiter
is extremely small. Your mass is attracted towards everybody
else's mass in your class, but again each force is very tiny.
In comparison to the Earth's gravitational attraction,
these very small forces are hardly worth considering.
The Earth's mass is approximately 6×10^{24} kg
(6 000 000 000 000 000 000 000 000 kg).
A 10 stone student's mass is 63.5 kg.

Deep space 5 N

Moon 80 N

e) How many kilograms are there in one stone?

Earth 500 N

The Earth's gravitational attraction from its huge mass creates
a **gravitational field** which pulls masses towards the centre of the Earth.
Any free moving masses will be **accelerated** downwards by the force.
This force of attraction is the object's **weight** on the Earth.
An object's mass might be constant wherever it is, but its weight
can change. The amount of weight an object has depends on its position.
It the object is near a strong gravitational field, it will have a much greater
weight than if it were placed in a weaker gravitational field.
An object's **weight** depends on the strength of the **gravitational field** it is in.

Jupiter 1250 N

Equation 11 to learn

Weight = **mass** × **g (gravitational field strength)**	
newton, (N) kilogram, kg newton per kilogram, N/kg	

Sun 13 700 N

Position of student	Student's mass	Gravitational field strength (g)	Student's weight
Deep space	50 kg	0.1 N/kg	50 × 0.1 = 5 N
Moon	50 kg	1.6 N/kg (g_{moon})	50 × 1.6 = 80 N
Earth	50 kg	**10 N/kg** (g_{earth})	50 × 10 – 500 N
Jupiter	50 kg	25 N/kg ($g_{jupiter}$)	50 × 25 = 1250 N
Sun	50 kg	274 N/kg (g_{sun})	50 × 274 = 13 700 N

Remind yourself!

1 Copy and complete:

Mass and ...... are not the same.
M...... is matter, the amount of atoms and
molecules inside something that make it up.
Weight is the ...... on a mass due to gravity.
The Earth, the Sun and the Moon have ......
fields which attract all mass.
The greater the gravitational ...... strength, the
more ...... an object will have inside that field.

Equation 11:
Weight = ... × gravitational field ......

2 One stone is 6.35 kg.
The Earth's gravitational field strength is 10 N/kg.
Calculate your weight using this information.
If you prefer, calculate the weight of a 40 stone
horse.

3 The Moon's gravitational field strength is
approximately 6 times less than the Earth's. If six
students could lift a Mini on the Earth, how many
would it take on the moon?

▶▶▶ 9d Turning forces and stretching forces

Turning forces

Have you ever tried to open a door by pushing near to the hinge?
It opens but it is much more difficult than normal.
If you have ever had to undo a wheel nut on a car, you will understand
the importance of having enough leverage.
You need a large **turning force** (or **moment**).

turning force (moment)	=	force	×	perpendicular distance to pivot
newton metres (Nm)		newtons (N)		metres (m)

Example 1 Opening a door

Force to open door = 40 N

Distance from handle to door hinge = 50 cm = 0.5 m

turning force = force × perpendicular distance to pivot
= 40 N × 0.5 m
= 20 Nm

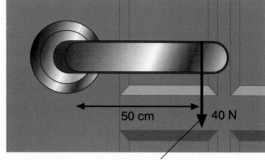

perpendicular force
(at right angles)

Calculate the turning forces for the following:

a) A spanner undoing a nut with force of 50 N applied 20 cm (0.2 m) from nut?

b) A tin opener being squeezed to pierce can with a force of 20 N, 10 cm from blade?

c) A wheel barrow lifted with a force of 150 N, 1 m from the wheel?

If you help a young child ride on a playground see-saw,
you need to use the principle of balanced turning forces.

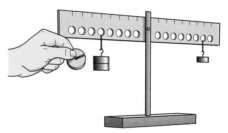

Sum of clockwise moments = sum of anti-clockwise moments.

If you sat at one end and you put the child at the other end, it is quite likely
that the see-saw would not be balanced. You would crash down,
sending the small child into orbit from the other end!

d) How would you adjust where you sat to make the see-saw balance?

When you move nearer the pivot, this reduces your **moment**.
It makes the anti-clockwise moment balance the clockwise moment.

e) Copy and complete this table of results. The beam must be balanced each time.

Force	Distance to pivot	Anti-clockwise moment	Force	Distance to pivot	Clockwise moment
5 N	6 cm	5 × 6 cm = 30 Ncm	10 N	3 cm	
2 N	12 cm		4 N		24 Ncm
12 N		36 Ncm	6 N		

Stretching forces and extension

When we stretch an object, it gets longer (extended).
If stretched enough, the object will **deform**. It has passed
its yield point. When the stretching force is removed,
the object will not return to its original shape or length.
If we stretch the object enough, eventually it will break.

> **f)** List three stretchy objects and three which would easily break if stretched.

Some materials are much easier to stretch than others.
Elastic bands, bungee rubbers and springs are all easily stretched.
We often use them to store strain energy. Push-along clockwork toys store
energy in stretched springs to give them kinetic energy when released.

> The extension of a material is **directly proportional** to the force stretching it.
> The increase in length per Newton is the same until the object gets **deformed**
> (permanent shape change).

Look at the graph on the right:
It shows the extension of a spring as the load
on it increases. The graph for an elastic band
or a bungee rubber would be very similar.

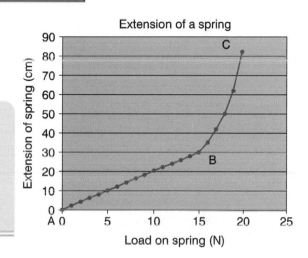

Extension of a spring

> **g)** Look at the graph between points A and B.
> How far does the spring stretch with each 1 N added?
>
> **h)** What happens to the extension per Newton
> from point B to point C?
>
> **i)** What do you think would happen
> to the spring if it was loaded beyond 20 N?

Up to the 15 N point, the spring is getting
2 cm longer for every 1 N of extra load added.
The extension is **directly proportional** to the load placed on the spring.
From 15 N to 20 N the extension gets much greater for each extra load.
The spring has passed its yield point. If the load was removed, the spring would
be longer than it was originally. Load the spring beyond 20 N and it would eventually break!

Remind yourself!

1 Copy and complete:

The turning force or ...... is equal to the force $\times$
the ...... from the pivot.
For balanced moments:
Clockwise moments = ... clockwise moments
The extension of a material is directly ...... to the
s...... force. If stretched beyond the yield ......
permanent deformation or breakage occurs.

2 Explain why, in terms of moments:
  i) It is easier to undo car wheel nuts using a
     long lever.

  ii) Keys have a wider handle than barrel.

3 Sketch a graph similar to the one above. Add a
line for a spring with three times the extension of
the one shown. Its yield point happens at
12 newtons.

Summary

Forces make things happen. The unit of force is the newton.
There are many different types of force:
friction, pushing and pulling, attraction and repulsion, weight, upthrust,
bending, twisting, tearing, squashing and stretching.

When forces are exerted; every **action** has an **equal** and **opposite reaction**.
*For example: when you stand on the floor, your weight pushes down but
the floor pushes you back up with an opposite force that equals your weight.*

Newton's first law refers to **balanced** and **unbalanced** forces:

An object that is moving at a constant speed will keep moving.   *Balanced forces*

An object that is stationary will stay stationary,   *Balanced forces*
until another force is applied to the object.

When forces are unbalanced, a **resultant** force acts.

Unbalanced forces will:

* make a stationary mass accelerate in the direction of the resultant force,
* accelerate (increase its speed) a mass if the resultant force
  is in the same direction as the original motion of the mass,
* decelerate (decrease its speed) a mass if the resultant force
  is in the opposite direction to the motion of the mass.

Mass is **not** the same as weight.
The amount of mass an object has is the same, wherever it is.
Weight is the force due to gravity on mass.
An object's weight depends on the strength of the **gravitational field** it is in.

Equation 11

weight = mass × g (gravitational field strength)
newton, (N)   kilogram, kg   newton per kilogram, N/kg

For example a student on the Earth:

Student's mass	g_e Gravitational field strength	Student's weight
50 kg	10 N/kg	50 × 10 = 500 N

Turning forces are often called **moments**. Their units are newton metres (Nm),
turning force (moment) = force × perpendicular distance to pivot.
When the turning forces on an object are balanced:
Sum of clockwise moments = sum of anti-clockwise moments.

The extension of a material is directly proportional to the stretching
force applied to it. The increase in length per newton is constant until
the object gets deformed (permanently changes its shape).

Questions

1 Copy and complete:

i) The unit of force is the ......
A few examples of different types of force are ......, ..., ..., and ......

ii) A force exerted by an object on a surface gets an equal and ...... reaction from the surface.

iii) Newton's ...... law refers to the effects of balanced and ...... forces.
When unbalanced ...... act on an object, it will start moving, ... up, slow...... or stop.

iv) Mass and ...... are not the same.
Weight is the force due to ...... on mass.
The weight of an object depends upon the strength of the ...... field it is in.
Weight = ... × gravitational field ......
The gravitational field strength on the ...... is 10 N/kg. A mass of ... kg weighs 500 N

v) Turning forces are often called ......
Turning force = force × ...... from pivot.
For balanced turning forces:
...... of clockwise moments = sum of ...... wise moments.

vi) When a ...... is stretched, the amount it extends by is directly ...... to the stretching force applied to it.

2 Construct a spider diagram or draw a poster to define different types of force.
For each force, either draw or write an example of where you can see it in action.
For example:
Twisting – opening a drink bottle.
Pulling – opening a door.

3 Explain each of the following, using ideas about balanced and unbalanced forces.

i) A book on a desk stays stationary.

ii) You can jump higher on the Moon than you can on the Earth.

iii) A car parked on soft ground can sink in.

iv) When you move a cup full of drink quickly, the drink can spill over the side.

4 Calculate the following weights for a student's CD player, its mass is 2 kg.
Weight = mass × g (gravitational field strength)

i) Its weight on Earth (g_e = 10 N/kg)

ii) Its weight on the Moon (g_m = 1.6 N/kg)

iii) Its weight on Neptune (g_n = 14 N/kg)

iv) Its weight on the Sun (g_s = 274 N/kg)

5 Equation 3 from page 48 is used to calculate work done upwards:
Change in gravitational
potential energy (PE) = weight × height
This equation can be combined with equation 12 to give:
Change in gravitational
    potential energy = mass × g × height.

i) Find the energy stored in a 3 kg box on 2 m high shelf. (g_e = 10 N/kg)

ii) Find the potential energy of the CD player in question 4 if it was held 4 m above the surface each time.

6 I) What would be the moment for a 5 N force applied 6 m from a pivot?

ii) What distance from a pivot would give a turning force of 36 Nm from a force of 12 N?

iii) To balance the beam below, what force would A need to be?

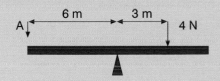

7 If 1 N force on a spring makes it stretch by 2 cm, find the stretch for these loads:

i) 4 N, ii) 7 N, iii) 15 N, iv) 0.5 N

Assume that none of the loads make the spring deform. (It always returns to its original length when it is unloaded.)

CHAPTER 10

Speed, velocity and acceleration

▶▶▶ **10a Speed**

Which goes faster on a skate board, a kangaroo or a dolphin?
You could push them down a hill and see which one makes
it to the bottom first. Mind you, the dolphin would probably slide off
on the first bend while the kangaroo might decide it's quicker to hop!
The idea of this race is very silly, but you still could predict a winner.
You might decide the kangaroo will win because it can kick better,
or maybe the dolphin would win because it is more streamlined.
Either way, you know what we mean by the fastest.
You understand speed.

a) Which do you think would win the race, assuming they
were both approximately the same mass? Why?

The speed equation

Equation 12 to learn

$$\text{Average } \textbf{speed} = \frac{\textbf{distance travelled}}{\textbf{time taken}} \quad \begin{array}{l}\text{(metres, m)}\\[4pt]\text{(seconds, s)}\end{array}$$

The clever thing about the speed equation is that it is easy to remember.
If you were asked the fastest speed you have travelled at in a car
or on a bus, you would give an answer in miles per hour (mph).
Miles are used to measure distances.
Hours are used to measure time.
If you remember the per as divided by, you know that equation:
Miles **per** hour (mph) ⟶ Miles **divided by** hours ⟶ Distance **divided by** time

The main unit of speed is the metre per second (m/s).

*Often the symbol / is used instead of writing **per** but it means
the same thing.*

Any unit of distance divided by any unit of time gives a unit of speed.

b) Copy and complete the table opposite:

Units of distance and time		Unit of speed
km	h	
cm	year	
mm	hour	
		mph

The speed equation triangle

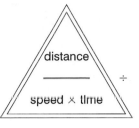

Example: a hungry student

A student is running down the corridor to get to the front of the lunch queue.
The corridor is 60 m long. It takes him 12 seconds to run down it.
Distance travelled = 60 m
Time taken = 12 s
Using:

$$\text{Average speed (m/s)} = \frac{\text{distance travelled (metres, m)}}{\text{time taken (seconds, s)}}$$

His average speed = 60 m ÷ 12 s
                  = 5 m/s

---

c) A ten-pin bowling ball travels in a straight line (well nearly) along a 10 m lane in 2.5 seconds. What is the ball's average speed?

d) A toddler has a go at skittles, scoring a strike with her first bowl. If the bowl's speed is 1 m/s, how long does it take to hit the skittles?

Average speed

When travelling, you very rarely travel at your average speed for the whole journey! How many times have you finished one lesson and then had to move **instantly** to the other side of the school to be in time for the next lesson?
Your **average speed** between the two parts of the school will be **less** than your **maximum speed** when running.
When you are queuing to go down stairs, your speed will be much less than your average speed. Of course, you're never fast enough for the teacher!

Remind yourself!

1 Copy and complete:

The average ......of a moving object is found by using equation 12:
...... = ...travelled ÷ ......taken
The main unit for measuring speed is the metre per ...... (......).
Two other common units used in speed measurement are ......per ...... (kmph) and miles per hour (......).

2 In question c) you found various other units for speed. Use them to pick the most suitable unit for each of these.

i)   A flying aeroplane.

ii) A suspension bridge expanding on a hot day.

iii) England and America moving apart.

3 Find out about the unit called a light year. Is it a unit of speed?

Look at the cartoon below.

This cartoon shows some students queuing for their lunch.
Egbert wants to be Ermintrude's new boyfriend so he is trying to jump
the queue to join her. Meanwhile Ermintrude waits her turn patiently,
standing about 2 m from the hatch.
The cartoon clearly shows how Egbert gets to '*Ermy*' (his nickname for her).
Without the cartoon you would need far more words to describe
what happened in the queue.

a) Try to explain how both Egbert and Ermintrude move relative
to the lunch hatch in only two short sentences.

Another way to show how both Egbert and Ermintrude move
is on a **distance–time graph**.
The graph below shows Ermintrude's motion as a flat line.
This is because she stayed 2 m away from the hatch for the full 10 seconds.
The line for Egbert's motion
has level and sloped sections.
This is because he started
off 6 m away from the hatch
but eventually he was 3 m
away. The level sections
show where he was stationary.

b) What motion do you think
the sloped sections show?

c) How far away from the hatch
was he after 5 seconds?

d) How far away from '*Ermy*' was
he after 8 seconds?

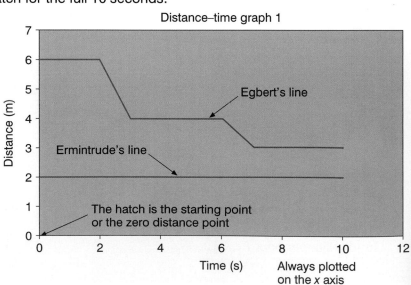

Distance–time graph 1

A **distance–time graph** shows how an object moves. These graphs show the position and speed of the object in relation to a fixed point (often its starting point). The distance and direction that an object is away from the fixed point is called its **displacement**. The graph on the right shows an object that is stationary. Its displacement away from the starting point is 2 m

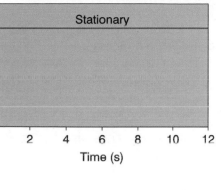

Distance–time graph 2

> A stationary object is shown by a level line on a distance–time graph.

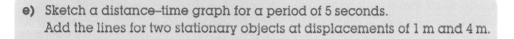

e) Sketch a distance–time graph for a period of 5 seconds. Add the lines for two stationary objects at displacements of 1 m and 4 m.

A moving object is shown on a distance–time graph as a line which is either going up or going down.

> Lines going up show increasing displacement away from the starting point.

> Lines going down show decreasing displacement, closer to starting point.

The object is moving further away from its starting point.

The object is coming back towards its starting point.

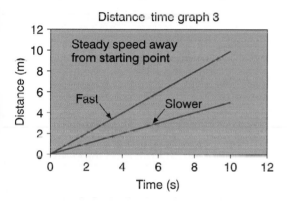

Distance–time graph 3

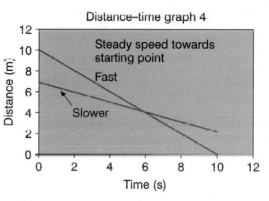

Distance–time graph 4

The steepness of the line on a distance–time graph shows the speed of the object.

> A steep line (large gradient) up or down shows a fast moving object.
> A line with a gentle slope (low gradient) shows low speed.

Remind yourself!

1 Copy and complete:

A distance–...... graph can be used to represent the ...... of an object. The distance travelled or d...... away from the starting point is always plotted on the *y* axis. ...... is always plotted on the *x* axis. The gradient of the line gives the ...... of the object. Flat line for ...... objects, a steep line for ...... objects.

2 Look at this distance–time graph for someone moving.

Distance–time graph 5

i)   At what points are they stationary?

ii) When are they moving the fastest?

iii) What displacement do they finish with?

3 Suggest how an accelerating object would appear on a distance–time graph?

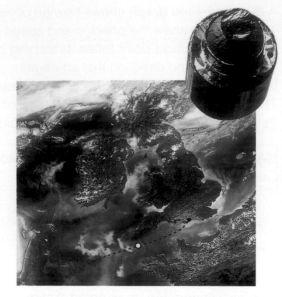

Ships have crossed the Atlantic for hundreds of years. They can travel for days and find their port with few changes to their course. Modern ships use high-tech satellite navigation systems to give them an exact location at any time. Two hundred years ago the only navigation aids were compasses, clocks and the stars. They also had ropes with big knots tied in them. Sailors used the ropes to work out their speed through the water. Having a clear idea of speed was not enough for the sailors to find their way. To navigate properly at sea, you must know your direction as well as your speed. The compasses, clocks and star maps helped navigators to work out their velocity.

a) What is the unit used at sea to measure speed?

Ships use satellites to find their exact position at sea.

Speed and direction

Two coaches are travelling at 60 mph along the M6 motorway.
They are 30 miles away from the centre of Birmingham.
The first coach has a puncture so its driver telephones the second coach to ask for help. The second coach is empty so its driver promises to collect the stranded passengers from the first coach.
Even though the driver of the second coach does not slow down or get lost, it still takes him an hour to get to the broken down coach.

b) Why does it take the second coach an hour to get to the first coach?

If both coaches are travelling towards Birmingham at the same speed, you would imagine that the second coach would only have to stop by the first one. Maybe it would have had to turn around at the next junction if it had already passed the first coach.
In fact, both coaches were 30 miles away from Birmingham but travelling in the opposite direction on the M6. The first coach from London was heading north-west, while the second coach from Liverpool was heading south-east.
We needed more information than just the coaches' speed to solve this problem!

For each of the following moving objects, decide if more than just its speed is required to describe its motion.

c) Two racing cars on the home straight, fighting for first place.

d) An aeroplane crossing Europe travelling at 400 mph.

e) Ermintrude running away from Egbert.

f) A ship crossing the Channel to France.

Velocity

An object's speed gives no information about where it is going.

> **Velocity** defines the **speed** and **direction** of a moving object.

The calculation for velocity is the same as that for speed:

Equation 12 (to find velocity in this case)

$$\text{Average velocity (m/s)} = \frac{\text{distance travelled (metres, m)}}{\text{time taken (seconds, s)}}$$

For a velocity, the direction of the object must also be given.

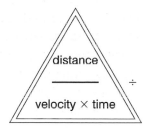

distance ÷ velocity × time

North, east, up, down or even backwards, these are all directions that can be given with a speed to make it a velocity. Normally a more exact direction is chosen by deciding the **positive direction**. Anything moving in that direction would have a positive velocity. Any objects moving in the opposite direction would have a negative velocity.

15 m/s

−5 m/s

The bus is travelling in the positive direction (as chosen). The car has a negative velocity because it is travelling in the opposite direction to the bus.

Example: friends meeting in London

Two friends are travelling to London to go on the London Eye.
Shanti lives in Bedford which is north of London.
It takes her 2 hours to travel the 50 miles by train and tube.
Rubi lives in Croydon in South London. She takes 1 hour to do her journey by bus, travelling approximately 10 miles.
To find the girls' velocities, taking north as the positive direction:

Shanti	Speed	50 miles ÷ 2 hours	= 25 mph
	Her direction is south so her velocity is		= −25 mph
Rubi	Speed	10 miles ÷ 1 hour	= 10 mph
	Her direction is north so her velocity is		= 10 mph (+10 mph)

Remind yourself!

1 Copy and complete:

Velocity is speed with ......
The velocity of a moving object has the same value as its ...... so it is calculated using the same ......
For a velocity, the direction an ...... is moving has to be clearly shown as well.
Velocities can be ...... or negative, once a positive direction has been selected.

2 When you are on a roller coaster ride at a theme park, your speed around a long bend might be far too fast but constant.
How would your velocity be changing while you went around the curve?

3 Aeroplanes can fly along the same flight path in opposite directions without colliding. As well as velocity, what other information is used to describe their motion?

▶▶▶ 10d Velocity–time graphs

Do you catch a bus to school? A typical bus ride will include a number of stops. Each time the bus pulls away from a stop it **accelerates** up to a constant cruising velocity (probably walking pace if you live in a city). When the bus approaches its next stop, it **decelerates**, reducing its velocity until it has stopped again. After all the other stops, eventually you get off at your school, only to do the reverse journey on the way home.

(Acceleration and deceleration are described in more detail on pages 138–139)

A **velocity–time graph** is a useful way of showing changes in velocity.

Look carefully at the velocity–time graph below:

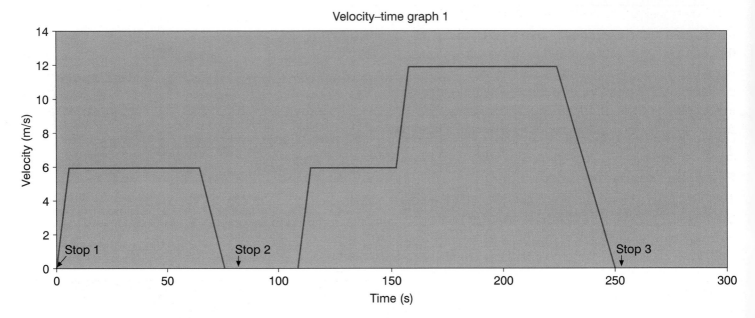

Velocity–time graph 1

This graph shows how the velocity of the bus changes as it travels through some city traffic. When the bus leaves Stop 1 it accelerates up to the speed of the rest of the traffic. It then travels with a constant velocity of 6 m/s (approximately 14 mph) until it has to brake for passengers at Stop 2.

The graph and its description above should help you answer these questions. On a velocity–time graph, what is the shape of the line for:

a) Constant acceleration?   **b)** Constant velocity?   **c)** Deceleration?

d) Look at the graph between bus Stops 2 and 3.

What happens to the flow of traffic at approximately 150 s?

Reading velocity–time graphs

When an object is stationary, it has no velocity.
The brown line on the graph opposite is a plot
of a stationary object. All of its points are on the x axis
where y − 0, so it has zero velocity.
The other two lines on the graph show constant velocity.
The red line shows a constant velocity of 3 m/s.

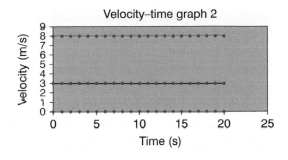

Velocity–time graph 2

> Constant velocity is shown as a level line on a velocity–time graph.

e) What velocity is shown by the blue line?

Acceleration is change in velocity.
Positive acceleration is when an object's velocity increases.
Deceleration (negative acceleration) is when the velocity
of a moving object decreases.

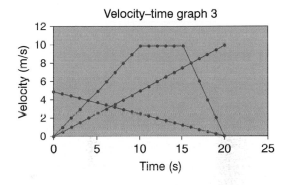

Velocity–time graph 3

> The gradient of a velocity–time graph shows acceleration.
> The steeper the line, the greater the acceleration.

Constant positive acceleration is shown by a straight line
climbing. Deceleration is shown on a velocity–time graph
by a straight line going down.

The graph below shows a velocity–time graph for an aeroplane
flying from London to Edinburgh. Its return journey back to London
is shown on the second half of the graph.

Velocity–time graph 4

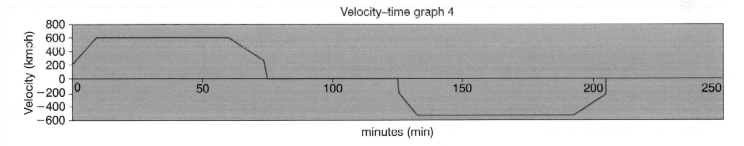

minutes (min)

f) Why is the second half of the graph below the x axis?

Remind yourself!

1 Copy and complete:

Velocity–...... graphs show how the ...... of a
moving object varies with time.
A ...... object has no velocity, giving a level line
along the ...-axis. Constant velocity gives a ......
line at the value required.
Constant change in velocity produces a steady
gradient, ... for acceleration, down for d......
The steeper the line, the greater the ......

2 Sketch the shape of a velocity–time graph for a
shopping trip. The journey involves walking for
15 minutes, taking a train for 30 minutes and
then wandering around clothes shops for
3 hours.

3 What type of movement does a curved line
represent on a velocity–time graph?

Have you ever found yourself falling from the top
of a sheer drop? If you have been on a roller coaster
you will know the feeling. You watch all the people in the queue
(for at least half an hour) as they have their turns. They scream
when they go around the *slightest* of bends and down *tiny* slopes.
You think that they must be all very cowardly, until suddenly
its your turn. Those gentle bends seem to jolt you right out of your
seat while the tiny slopes turn into free falls that seem
to last forever. Then the roller coaster starts to go backwards
before it finishes with a couple of twists upside-down!

a) What types of forces do you feel on a roller coaster?

Your body is being given a series of large **unbalanced forces**.
Every force makes you change your speed in the direction of the force.
You are continually being **accelerated** in all sorts of different directions.
You might be one of those people who gets a buzz from the forces
you experience on the most extreme rides (or maybe you are more
used to looking at your lunch *again* afterwards!).

Measuring acceleration

The acceleration of an object tells us how much its velocity changes each second.

Velocity is measured in metres per second (m/s).
Acceleration is the change in velocity every second.
Its units are metres per second per second (metres per second squared) (m/s^2).

An unbalanced force on an object makes it accelerate in the direction of the force.
The greater the force, the greater the acceleration it produces.

b) How do you think the mass of an object affects the acceleration
it gets from a force?

If you have ever helped push a car, you will know that the more
massive an object is, the harder it is to make it accelerate.

c) Which will accelerate more with a 50 N kicking force:
a 0.1 kg tennis ball or a 5 kg bowling ball?

Calculating acceleration

Equation 13 to learn

$$\text{acceleration} = \frac{\textbf{change in velocity}}{\textbf{time taken for change}} \quad \begin{array}{l}\text{metres per second (m/s)}\\ \text{seconds (s)}\end{array}$$

metres per second squared (m/s²)

This equation is often used in this symbol form:

$$a = \frac{(v - u)}{t}$$

where a = acceleration
v = final velocity
u = original velocity
t = time taken for change

You can write the symbol form like this $a = (v - u) \div t$

d) Explain how change in velocity can be found from $(v - u)$?

Example 1: the Mini and the Volvo Turbo

A Mini and a Volvo turbo set off from a set of traffic lights together.
After 10 seconds, the Volvo gets up to 30 m/s but the Mini takes an extra 5 seconds.
Find both cars' accelerations:
Using the symbol formula above in the form $a = (v - u) \div t$;

The Mini: Original velocity u = 0 m/s
        Final velocity v = 30 m/s
        Time taken t = 15 s
        $a = (30 - 0) \div 15$ s
        = 2 m/s² (SLOWER)

The Volvo: Original velocity u = 0 m/s
        Final velocity v = 30 m/s
        Time taken t = 10 s
        $a = (30 - 0) \div 10$ s
        = 3 m/s² (FASTER)

e) At the lights in the example above, there was also a 1000 cc motor bike which got to 30 m/s in only 6 seconds. Calculate its acceleration.

Example 2: a train

The train is travelling at 60 m/s and it takes 2 minutes to slow down and stop at a station. What is its acceleration?
u = 60 m/s, v = 0 m/s, t = 120 s (60 sec in a minute)
Change in velocity $(v - u) = (0 - 60) = -60$ m/s
Finding acceleration using $a = (v - u) \div t$; -60 m/s $\div$ 120 s $= -0.5$ m/s²
The train has a negative acceleration because it is slowing down (decelerating).

Remind yourself!

1 Copy and complete:

...... is how much velocity changes in a ......
Whenever an unbalanced ...... is applied to an object it will accelerate.
How much an object accelerates depends on its ...... and the ...... of the force applied.

Equation 13:

$$\text{acceleration} = \frac{\text{change in ......} \quad \text{(m/s)}}{\text{...... taken for change (s)}}$$
m/s²

2 The space shuttle decelerates by sending a rocket burst out in the opposite direction to the one it is travelling in. What would happen if the rocket was used for too long?

3 Find out about Newton's second law and how it links force, mass and acceleration together.

Summary

Equation 12

$$Average \text{ speed} \text{ (m/s)} = \frac{\text{distance travelled (metres, m)}}{\text{time taken (seconds, s)}}$$

Speed is measured in metres per second (m/s).

Distance–time graphs show the position and speed of an object from a fixed point.
The distance and direction of an object from the fixed point is called its **displacement**.
A stationary object is shown by a level line.
Lines going up show an object getting further away.
Lines going down show an object getting closer.
A steep line (large gradient) up or down shows a fast moving object.

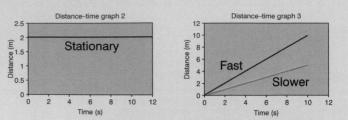

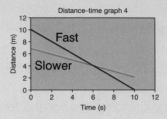

Velocity tells us the **speed** and **direction** of a moving object.
To calculate a velocity, the speed equation is used.
We have to decide which direction is positive because velocity can be positive or negative.

Velocity–time graphs
We show constant velocity as a level line on a velocity–time graph.
Acceleration is change in velocity. Positive acceleration is when an object's velocity increases. Deceleration (negative acceleration) is when the velocity of a moving object decreases.
The gradient of a velocity–time graph shows acceleration.
The steeper the line, the greater the acceleration.
Constant positive acceleration is shown by a straight line climbing.
Deceleration is shown on a velocity–time graph by a straight line going down.

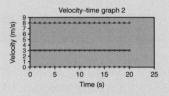

The acceleration of an object tells us how much its velocity changes each second.
Unbalanced forces on an object make it accelerate in the direction of the resultant force.

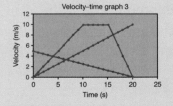

Equation 13

$$\text{Acceleration} = \frac{\text{change in velocity}}{\text{time taken for change}} \quad \begin{array}{l}\text{metres per second (m/s)} \\ \text{seconds (s)}\end{array}$$

metres per second squared (m/s²)

This equation is often used in this symbol form:

$$a = \frac{(v - u)}{t}$$

where a = acceleration
v = final velocity
u = original velocity
t = time taken for change

You can write the symbol form like this $a = (v - u) \div t$

Questions

1 Copy and complete:

i) Equation for average speed:

...... = distance ...... ÷ time ......

Speed is measured in ...... per second (m/s).

ii) A distance–...... graph can show the position of an ...... relative to a fixed point. A level line represents a ...... object. Lines going up or down represent ...... objects.

iii) Velocity is speed and ...... Velocity is calculated using the same equation as ...... A ...... for the moving object must be given.

iv) On a velocity–...... graph, constant velocity gives a ...... line. ...... is shown by a line going up and ...... by a line going down.

v) Deceleration is negative acceleration. When ...... balanced forces are applied to an object, it will accelerate in the direction of the ...... force.

vi) The equation for calculating acceleration:

...... = change in ...... ÷ time taken.

Symbol form: $a = (... - u) ÷ ...$
where; v = final speed, u = original speed and t = time taken.

2 Copy and complete this table:

Object	Distance (m)	Time (s)	Speed (m/s)
Y10 boy	20	20	
Milk float		50	4
Racing flea	12		6
Helicopter	10 000		40
Space liner		60	5000

3 Describe the motion of the students represented in this distance–time graph.

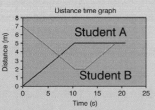

4 Sketch a distance–time graph to show a student getting up from his chair. He goes to the TV and then pauses to check some channels. Finally he runs back to his chair when he finds something worth watching. (The remote is broken.)

5 Two girls are on their way to meet for a Saturday shopping trip. Cassie has a velocity of +3 m/s and she is 54 m away from the coffee bar meeting point. Hannah is only 36 m away from the coffee bar, her velocity is –2 m/s.

i) Calculate which girl gets to the coffee bar first.

ii) If they both had a positive velocity, would they meet before they got to the coffee bar?

6 The velocity–time graph below shows the motion of two skateboarders.

i) Describe the motion of each skateboarder.

ii) Find the constant velocity of skateboarder 2.

iii) The area under a velocity–time graph is the distance travelled. Which skateboarder travelled the furthest?

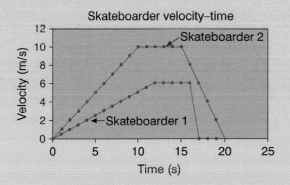

7 i) Find the acceleration of a racing car that goes from rest to 84 m/s in 12 seconds.

ii) If the car slows to 36 m/s in the next 8 seconds, what was its acceleration?

CHAPTER 11

FRICTION and MOVEMENT

▶▶▶ 11a Friction and heat

Imagine if there was no friction for a day:

Good things:

  You wouldn't be able to write, your pen would slide over the page.

  You wouldn't be able to push keys on a computer.

  You couldn't even get to school, all the roads would be too slippery.

Bad things:

  You'd be very hungry as you wouldn't be able to pick up your food.

  Mobile phones would be flying everywhere like wet bananas.

  You might even keep sliding out of bed every time you moved.

a) List a few more good and bad things about a life without friction.

You need friction to hold objects.

Friction

Have you ever tried to stand up on ice? It's easy until you try to move. Move too quickly and 'the only way is down'. Ice is a low friction surface. Friction is a contact-based force, acting where surfaces touch.

Friction opposes (resists) motion between surfaces.

If two surfaces are in contact, the frictional forces will try to keep both surfaces stationary, stopping either from moving.

If both of the surfaces are very smooth, movement will be easy as little friction will oppose motion. If the surfaces are very rough and able to 'dig into' each other, there will be high friction between the two surfaces.

A large force will be needed to move one surface over the other.

The friction between most surfaces is somewhere between these two extremes.

b) Think of two surfaces that would have high friction and two with low friction.

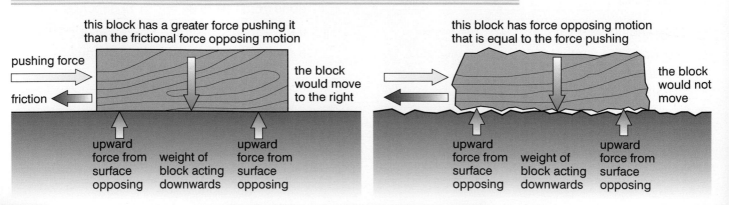

Friction and heat

Wherever you find friction between surfaces, you also get wear.
Shoes slowly wear their soles out while carpets get worn down
and linos get polished. The particles at the surface of materials
in contact with other materials get dislodged with the constant rubbing.

c) What happens to the temperature of two surfaces rubbed together?

Heat energy is always being released when friction occurs.
The temperature change on a shoe sole while walking would be very difficult
to measure but you could measure a car tyre's temperature change.
Too much heat produced by friction can be very destructive.
Oil and grease are used to **lubricate** all sorts of machinery.
Car engines, bike wheels, washing machine drums and hair driers
all rely on lubricated bearings to support moving parts. If the oil or grease
dries up, the heat from increased friction often causes instant damage.

The amount of heat energy released when friction occurs depends
on the work done in overcoming friction. Work done was discussed on page 46.

Equation 2			
	Work done =	**Force applied** ×	**distance moved**
	or energy transferred		in the direction of the force
Units	joule (J)	newton (N)	metre (m)

Example: the heat produced by rubbing hands

You rub your hands together to keep them warm.
Over 10 seconds you rub them 20 times backwards and forwards.

Force to overcome friction = 2 N

Distance, once backwards and forwards = 0.30 m (30 cm)

Distance travelled in total (0.30 × 20 = 6 m) = 6 m

Total work done against friction = 2 N × 6 m

= 12 J

A block of wood weighing 20 N is moved through 2 m across a path.
It is being pushed by a force of 12 N. The friction resisting motion is 10 N.

d) Sketch the block showing all the forces acting on it.

e) Calculate the work done against friction.

Remind yourself!

1 Copy and complete:

...... is a contact-based force. It happens whenever two surfaces rub together.
Friction always ...... motion. It tries to stop the two surfaces in contact from ...... past each other. When friction occurs, ... is produced.
L...... in bearings and joints stops wear by drastically reducing ......

2 Write a brief plan for a method of comparing the temperature changes produced on a trainer's sole after running. The trainer could be tried on a selection of different surfaces.

3 Why do snooker and pool players always chalk their cues before they hit the cue ball?

Imagine this:

You are living on the Moon. It's race time!
You have just set off and are ahead in your specially converted
oxygen tanked sports car when you are overtaken by a removal van.

The van's bigger engine and low ratio gears help it to carry loads
at a reasonable speed against opposing frictional forces.
On the Moon, there is no **wind resistance**. The huge advantage
of **aerodynamic** styling that the sports car has is completely useless.
The van would be away fast giving the sports car an unexpected challenge.
Mind you, it would probably skid off the road at the first bend.

a) How does increasing the weight in a box affect the friction
that must be overcome to drag it along the floor?

Friction in fluids

Friction in fluids (a liquid or a gas) is sometimes called **drag**.

When an object moves through a fluid, the friction on it increases
as it moves faster.

If you have ever been swimming you will know about friction.
Olympic swimmers wear specially designed costumes to reduce
the drag on their bodies as they swim.
Some swimmers also wear **hydrodynamic** (low friction in water)
swimming hats while others have been known to shave their heads!

Using the idea of fluid friction, describe what happens when you:

b) Dived into a swimming pool properly.

c) Dived in but did a 'belly flop' instead.

Friction in fluids acts between the surface of the object
and the particles of the fluid touching it.
It is almost as if the particles close to the object become slightly
stuck while those further away still flow easily.
This creates **turbulence** within the fluid around the object
dragging it back (slowing it down). Ships, speed boats and yachts
are all designed so that the effects of fluid friction are reduced.

d) How does a hovercraft reduce its fluid friction with the sea?

Modern hydrodynamic swimwear.

Air resistance

Air is a mixture of gases so it is a fluid. The friction resisting the motion of an object travelling through air is called **air resistance**. Hold this text book flat and move it quickly across in front of you, without hitting anybody! Move it back just as quickly but this time hold it upright pointing away from you. You should feel more air resistance. You might notice some air flowing past you, pushed by the book. Vehicles are given aerodynamic shapes (**streamlined**) to ensure that they *cut* through the air with ease. Cars with low air resistance are more economical. Sports cars and racing cars have to be aerodynamic.

Notice the streamlined headlights on this car.

e) Sketch a sports car and also a lorry.
Show the air flowing over each vehicle as arrows.

The engine produces a **driving force** that pushes the car forwards. Friction from the road and mostly from air resistance opposes the driving force. When the car gets close to its maximum speed, the driving force and frictional forces will balance and cancel out. There will be no resultant force so the car will not be able to accelerate. Instead the car will just keep going at a steady speed.

weight

Terminal velocity

If you were to jump out of an aeroplane you would start to accelerate downwards because of the pull of the Earth's gravity. It would not be long before you reached **terminal velocity**. Your weight pulling you down is being opposed by the air resistance on your body. When these two forces balance, you will stop accelerating and just continue to fall at a constant speed, your terminal velocity. When you open your parachute, you slow down as you have increased your air resistance. Soon you will have a new, lower terminal velocity.

air resistance

Remind yourself!

1 Copy and complete:

The motion of an object in a ...... (a liquid or a gas) is opposed by ...... forces.
Fluid friction is often called ...... Friction to movement in the air is called ....
Air resistance can be reduced by making the shape of an object more ...... (streamlining it).
A vehicle travelling at a constant speed has its ...... force balanced by opposing frictional forces from the road and ......
A falling object will reach ...... velocity (constant speed) when its ...... equals its air resistance.

2 True or false?
When your parachute opens, you go back up again.
Discuss this idea with a friend.
Write an explanation of what you think happens.

3 How would the terminal velocity of a falling stone be affected if:

i)   It fell from the air into water?

ii) It was dropped into treacle?

iii) It was falling above the Moon?

▶▶▶ 11c Friction, brakes and stopping

If you have ever been in a car or on a bike during a skid, you will
know the importance of stopping safely afterwards!
Being inside a car while it is in an uncontrolled skid is excellent fun,
as long as you are safely on a special skid pan. Out on the road,
it is not the skid that is the problem, it is the things that you hit.

a) List three factors that affect how quickly a car stops.

Tyres

The type and condition of the tyres fitted to a car are very important.
The amount of friction between the tyre and the road depends
on how much of the tyre is touching the road and the type of road surface.
Racing cars have specially designed tyres to give them maximum grip
(high friction) when cornering and stopping. In dry weather, the race
tyres fitted are totally smooth (slicks). This gives maximum contact
between tyre and road surfaces. These tyres are made of a soft rubber
to increase their grip even further, causing them to wear out faster.
Because of this, racing cars have to change tyres at least once during a race.

'Racing slick' tyre.

Smooth tyres would be totally illegal on a road-going car. They might give
more grip in the dry but they would be highly dangerous in the wet.
Water would get trapped between the tyre and the road at the point
of contact. This would seriously reduce friction and cause skidding.
Legal tyres still have enough tread to channel water away from the contact
point between the tyre and the road. New tyres are better than old
ones because their tread is deep enough to channel the water away quicker.

b) What happens when a car aquaplanes?

c) Explain why *off road* tyres used by 4 × 4 vehicles have very deep treads.

Normal car tyres give good grip over most types of road surfaces.
They have to perform well in the rain, over mud and leaves during
autumn and in ice and snow during the winter.
People like their tyres to last for a long time as new tyres are expensive.

d) A tyre that lasts for many miles has a tread that doesn't wear down.
Which is better, tyres with a high grip and a shorter life or longer life tyres?

Skidding and brakes

The best quality tyres in the world would be useless on a car with poor brakes. The brakes on cars work by having a special high friction material which is rubbed against a smooth metal surface. The friction produced slows the wheel down. Slowing the wheel normally slows the car down too. Push the two surfaces together hard and the **braking force** is at its greatest.
Worn brake components give less, or sometimes virtually no, friction. This is why brakes on a car should be checked regularly to ensure that they are not too worn down.

A worn car disc brake pad next to a new one.

e) What happens when the braking force is greater than the friction between the tyre and the road?

Brakes stop the wheel, not the car.

If the friction between a tyre and the road is not enough, then the wheel will skid. The brakes will stop the wheel from turning at all rather than just slowing it down. Once a wheel starts to skid, it can be very difficult to get it to stop skidding and still keep braking. If one of your wheels starts to skid, the car can very quickly get out of control, skidding on into a crash. Taking your foot off the brake would stop the wheel from skidding for a while.

f) Would you stop braking when you are skidding towards another car *just in case it helped*?

Anti lock braking systems (ABS) work by *turning off* the brakes on the wheel that is skidding and then turning them on again. This is done very fast so the skid stops but the brake still operates. Each wheel has a speed sensor which sends information to a computer. If one wheel has stopped going round while the others are still rotating, it must have locked up. The brake on the skidding wheel is turned on and off until the car finally stops or the brakes all go off again.

Remind yourself!

1 Copy and complete:

...... between tyres and road surfaces gives a vehicle enough ...... to start, ... and turn. If a tyre is badly worn down it will be d...... on low friction surfaces (......, muddy or icy). The brakes on a car slow or stop the ...... from turning. If the friction on the road at the tyre is ...... than the braking force on the wheel, the wheel will ......

2 Before ABS were common, drivers were advised to *pump the brakes*, during a skid. This process involved the driver pushing and then releasing the brakes.
Do you think this process would help?

3 Find out how drivers in Sweden and Norway prepare their cars to drive in snow.

Did you know that it takes an oil tanker at sea about 5 miles to stop?
What about stopping on ice when you are hurrying into school?
You walk around a corner towards the main entrance and a herd
of elephants (some year 7 pupils) come rushing in the opposite direction.
You think a little and then you realize that you must stop.
Unfortunately before you know it, you have skidded on the icy path
and crashed, squashing five of them in the process.
The Deputy Head investigates but he can't decide if you were going too fast
for the *path conditions* or they were not walking with *due care and attention*.

A sports car and a large lorry are travelling next to each other on a motorway.
They are both travelling at 60 mph.

a) If they begin to stop using the same braking force, which will stop first?

b) When both vehicles have to stop in the same distance to avoid traffic,
which one will need to apply the largest braking force?

With the same stopping force, the car would stop quicker.
The lorry would have to apply a much larger braking force than the car
to stop in the same distance.

As a vehicle's speed increases, the **braking force** required
to stop it in a certain distance increases.
Stopping quickly when you travel faster requires a large braking force.

Stopping and kinetic energy

Kinetic energy is the amount of energy an object has because
of its motion. For more detail about kinetic energy, see pages 56–57.

Equation 8

$$\text{Kinetic energy} = \tfrac{1}{2} \times \text{mass} \times (\text{velocity})^2$$
Symbol form $\quad KE = \tfrac{1}{2}mv^2$

You can see from equation 8 that the amount of kinetic energy
a moving object has depends on both its mass and its speed (velocity).

The more kinetic energy an object has, the greater the braking force needed to stop it.

---

c) Which do you think would have more kinetic energy; a fast car or a slow bus?

d) Calculate the kinetic energy of a sports car travelling at 60 mph (27 m/s)
and a bus travelling at 20 mph (8 m/s).
Mass of car = 1000 kg, Mass of bus = 20 000 kg
(Hint: use equation 8, looking back to page 57 for a worked example.)

The energy transferred in stopping

You should have found in question d) that the bus had 640 000 J
of kinetic energy while the car had only 364 500 J.
The bus was moving three times slower but had twice as much energy!

Stopping a vehicle involves its brakes providing enough friction to give
a large enough braking force. The brakes provide a frictional force
for the whole distance that the car is stopping. They have to **do work**
to stop the car. This produces wasted heat energy as they slow it down.

e) Use the work done equation from page 46.
   i) Calculate the braking force required to stop the bus from question d) over 50 m.
   ii) Calculate the stopping distance for the car from question d) given a braking force of 6000 N.

Total stopping distances

Humans always like to think about a problem before they can solve it.
Imagine you are on the road riding your bike, and a car pulls out in front of you.
You have to decide, very quickly, what the best reaction should be.
Once you have thought about it for a bit, you probably decide to stop!
All the time you were thinking, you were still moving towards the car.

> Your **total braking distance** depends upon both:
> the distance you travel while you are **thinking** about stopping
> and the distance you travel once you have started to brake.

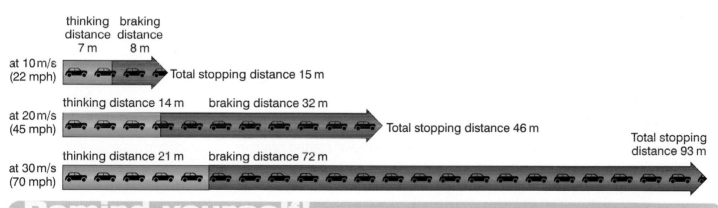

thinking distance 7 m   braking distance 8 m
at 10 m/s (22 mph)   Total stopping distance 15 m

thinking distance 14 m   braking distance 32 m
at 20 m/s (45 mph)   Total stopping distance 46 m

thinking distance 21 m   braking distance 72 m
at 30 m/s (70 mph)   Total stopping distance 93 m

Remind yourself!

1 Copy and complete:

As a moving object's ...... increases, the braking
...... required to bring it back to rest also
increases.
The greater the kinetic ...... a moving object has,
the larger the ...... force required to stop it.
Total ...... distance for a moving vehicle is the
...... distance (when the driver is considering
braking) and the braking ...... added together.

2 How would driving in heavy rain, fog or snow
affect:

i)   The stopping distance of a car?

ii) The driver's visibility?

3 Drivers who are tired, been drinking alcohol or
using drugs take longer to react to dangerous
situations.
Create an information booklet to encourage
drivers to be aware of the dangers and to be
more responsible.

Summary

Friction is a **contact** force, acting where surfaces touch.
Friction opposes (resists) motion between surfaces.
When friction produces too much heat, it can damage a machine.
Oil and grease are used to **lubricate** the bearings and joints
that support the moving parts in machines.
The amount of heat energy released as a result of friction depends
on the **work done** in overcoming friction.

Friction in fluids (liquids or gases) is sometimes called **drag**.
When an object moves through a fluid, the friction on it increases
as it moves faster. Olympic swimmers wear specially designed
hydrodynamic costumes to reduce the drag on their bodies.

Air resistance is the friction resisting the motion of an object travelling
through the air. We give vehicles aerodynamic shapes (**streamlined**)
to ensure that they *cut* through the air with ease.
Cars with low air resistance are more economical. They use less fuel.

Terminal velocity happens when objects fall.
Weight pulling an object down is opposed by the air resistance pushing
it back up again. The falling object will accelerate until these two forces
balance out. Then the object will stop accelerating and will continue
to fall at a constant speed. This is the object's terminal velocity.

Car tyres need to have good grip in the rain, mud, ice and snow.
Brakes stop the wheel, not the car. If there is not enough friction
between a tyre and the road, the wheel will skid.

As a vehicle's speed increases, the **braking force** required to stop
it in a certain distance increases. Stopping quickly when you travel
faster requires a large braking force as the vehicle has more kinetic energy.

Your **total stopping distance** is the distance you travel while you are thinking
about stopping plus the distance you travel once you have started braking.
Drivers who are tired or have been drinking alcohol or using drugs
take longer to react to a dangerous situation.
This means that their total stopping distance is increased.

Questions

1 Copy and complete:

i)   Friction is a ...... force.
It ...... motion between ......
Friction in moving machinery can cause damage as ...... is generated.
Oil and ...... are used to ...... bearings and joints in machinery.

ii) Friction in fluids is sometimes called ......
Air ...... slows objects ...... through the air.
Making their shape more aerodynamic or ...... reduces air resistance. Cars with low air resistance are more ......

iii) Terminal velocity occurs in ...... objects.
They travel at a ...... velocity once their weight acting ...... is balanced by air resistance acting ......

iv) Tyres need to have good ...... in all road conditions including ...... and ......
If the tyres grip is less than the ...... force, the wheel will ......

v)   The ...... a vehicle is moving, the more ...... energy it has. This means it will require a ...... braking force to stop it.

vi) The total ...... distance of a vehicle is the distance covered while ...... about stopping and the distance covered after ......

2 Describe how each of the following is only possible because of friction.

i)   Painting a wall.

ii) Fridge magnets sticking to a door.

iii) A drawing pin in a pin board.

iv) Writing with a pen.

v)   Kicking a football.

3 i)   Why do modern sprinters choose to wear body suits instead of traditional shorts and vests?

ii) Draw a sketch to show how a dragster racing car makes use of a parachute to stop. Label all the forces acting.

4 Decide if the following statements are true or false:

i)   A feather falls just as fast as a stone above the Moon's surface.

ii) A piece of paper has the same terminal velocity, whether it is a sheet or a tight ball.

iii) A coin dropped from the top of the Eiffel tower will have a greater terminal velocity than the same coin dropped from Big Ben.

iv) A parachutist would fall faster above the surface of the Moon than above the Earth.

v)   Washing a car makes it go faster.

5 Read through the following points and then decide if good tyres are really a good idea:
Tyres that grip better on the road normally wear out quicker. Replacement costs occur more frequently.
Good grip tyres are safer, especially in bad weather conditions, but accidents are normally caused by careless driving. Improve driving, and the tyres wouldn't need to be as good.
Good grip tyres cost more to make, use up better quality natural resources and make vehicles use more fuel.

6 Everybody knows that drinking and driving are very dangerous. Drivers react much slower and misjudge risks.
Driving while sleepy, not well, or while using drugs can be equally dangerous when things go wrong.
Use these points to make a hard hitting slogan to encourage people to take driving more seriously. THINKING DISTANCE!

CHAPTER 12

SPACE

▶▶▶ 12a The Earth and gravity

People have always been interested in Space.
For centuries it was believed that the Earth was the centre of the Universe.
If this was true, it would mean that the Sun and all the other planets
in our solar system would have to orbit the Earth.
People also believed that there was a roof over the world, a little like
an upturned colander. All the stars and constellations were
like tiny lights shining through the holes in the colander.

We now have a much clearer view of the Earth's place in the Universe,
thanks to the research carried out by the scientists Copernicus, Galileo
and Newton. Galileo even went on trial and spent the last years of his life
under house arrest. All because he dared to suggest that the Earth
went around the Sun! The church didn't agree.

a) How long does it take for the Earth to orbit the Sun?

b) How many hours does it take for the Earth to spin once on its axis?

c) If the Earth did not tilt on its axis, would we still have seasons?

The Earth, the Moon and the Sun

The Earth tilts at an angle of $23\frac{1}{2}°$. This is why we have seasons. When the UK has its summer, the northern hemisphere (top half of the Earth) is facing the Sun more than the southern hemisphere. Six months later, the opposite is true, we have winter and Australia has summer, as shown in the diagram below.

Our star, the Sun, provides the Earth with nearly all of its energy.

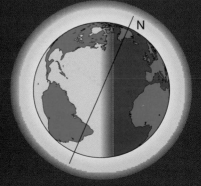

The Moon takes approximately 28 days to orbit the Earth. It is a **natural satellite** to the Earth. A full moon will be seen in the night sky. Sunlight is reflected off the whole visible surface of the Moon back to the Earth. A new Moon happens 14 days later. It is in the day sky and no sunlight is reflected back to the Earth.

The Earth rotates anticlockwise around the Sun. It takes $365\frac{1}{4}$ days or one year. Every four years we have a leap year with an extra day to make up the quarters. It keeps the seasons in the right part of the year!

The Earth spins taking one day or 24 hours to spin once. This gives us night on half the Earth and day on the other half. As we spin, both the Sun and the Moon appear to rise and fall (East to West). Really points on the Earth's surface are turning to face the Sun or the Moon, and then turning away as they move underneath.

Gravity and orbits

Gravity is the force of attraction between two masses. See page 124.
The mass of the Earth is approximately 80 times
greater than the mass of the Moon. The force of gravity
attracts these two huge masses together causing the lighter
Moon to orbit the Earth. The Moon is the Earth's **natural satellite**.
If it were possible to stop the Moon from moving forward in its orbit,
it would start travelling directly down to the Earth, crashing into the surface!
Fortunately this does not happen. The Moon's high speed forward
motion in its orbit keeps it *missing the Earth* as it falls.

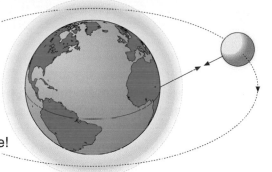

Gravitational attraction between the Earth and Moon.

The Sun's mass is nearly a million times greater than the mass of the Earth.
The strength of its gravitational field is enough to make all the planets,
asteroids and comets in our solar system orbit the Sun.
On the surface of the Sun, the gravitational field strength (g_s) is 274 N/kg.

> **d)** The Earth's gravitational field strength is 10 N/kg.
>
> How many times is the Sun's gravity greater than the gravity on the Earth?

> The gravitational field strength around a planet or star reduces with separation.
> The greater the distance from the planet or star, the lower the pull from its gravity.

In fact, as the separation doubles, the gravitational field strength decreases
by four times. Treble the separation and the field strength reduces
by nine times. Four times the separation, 16 times less field strength, etc.

> **e)** What is the pattern between increasing separation
> and decreasing field strength?

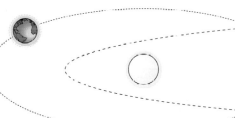

The planets orbit the
Sun with nearly circular
orbits a bit like squashed
circles or ellipses. The
Sun is very close to the
centre of their orbits.

Comets follow far flatter elliptical orbits,
with the Sun close to one end of their
orbit. They **speed up** as they 'loop' around
the Sun. The comets then travel out of
sight deep into the outer reaches of our
solar system before returning years later.

Remind yourself!

1 Copy and complete:

The Earth takes ...... days to orbit the Sun.
In one day or ... hours, the Earth spins once.
Gravity is the force that makes all the planets,
including the Earth, orbit the Sun.
Gravitational ...... gets weaker as you move
further away from their source planet or ......
Planets orbit the ...... with nearly circular elliptical
paths. Comets have ...... elliptical orbits.

2 How do we know that the Earth is not flat? What
if the Earth is really the centre of the Universe?
Write your own argument for or against these
points.

3 Find out:

i) Which planet has such an
elliptical path that it 'moves out
of order'?
ii) Which comet are we due to see next?

▶▶▶ 12b Satellites and orbits

Have you ever watched a news reporter in America talking
to a presenter in the UK. If you have, you probably noticed
a delay between the presenter asking a question and the reporter
in America answering. The delay happens because
the TV/radio wave has to travel a great distance.
Instead of just going from the UK to America, it travels thousands
of miles from the UK up to a satellite and then back down to America.

Satellites orbit above the Earth's surface but what keeps them up there?

An interview via satellite link.

Orbital motion

What happens when you throw a ball vertically upwards?
It comes straight back down and hits you on the head!
What if you throw the ball forwards as well as up?
This time the ball will travel across the ground before it hits.
What if you threw the ball so hard that it went for miles before it fell?
The ground below the ball would start to 'curve away' beneath.

The Earth is a sphere. If you travel above its surface in a straight line
far enough, the curve of the Earth would make you gain height.
You could almost think of a satellite in a low orbit continually falling
to the surface but missing as the surface keeps curving out of the way.

a) If you have a ball tied to a piece of string swinging around
above your head. What happens if you let go of the string?

Imagine swinging a ball on a string above your head. When you let go,
it flies off in a straight line. While it is going round, it behaves
a bit like a satellite in orbit. The string pulling on the ball is like the pull of gravity
on the satellite. If the gravitational field was suddenly removed, the satellite
would fly off into deep space in a straight line.

b) What do you think would happen to the satellite if the strength of the
gravitational field was suddenly increased instead?

The faster a satellite moves through space, the higher its orbit.
If it starts to slow down, the radius of its orbit will reduce.

A satellite's speed through space determines its height above the Earth.

Drag from the Earth's atmosphere will slow down any object that gets close.
If a satellite gets too close, it will either burn-up or fall down to the surface.

Different types of satellite

Some satellites have low orbits while others are much higher.
Those with low orbits travel slower through space, but
they do travel across the surface of the Earth quicker.
This is because the circumference of their orbit is much shorter
than those of the higher satellites.

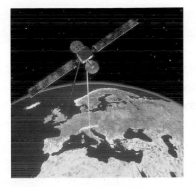

> Higher satellites take longer to go around the Earth.

c) A particular satellite takes two hours to cover the Earth's surface.
If it went around the Earth between the north and south poles, would the
rest of its route also cover the same ground on each pass?

Polar satellites are low level satellites. They cross
the surface of the Earth quickly. As the Earth spins
underneath, a different part is visible to them.
This means that they can be used to study the whole
of the Earth's surface over a series of passes.
This type of satellite is often used for military and other
monitoring purposes.

Geo-stationary satellites are positioned so that
they stay above the same part of the Earth's surface
all the time. To do this, they have to be above
the equator and orbit the Earth in 24 hours.
This is only possible from a height of 36 000 km.
As all geo-stationary satellites have to be at
the same height, there is only enough space for
around 400 of them. These satellites are used
for communications including TV and radio.

d) In the UK, which direction do the satellite dishes all point?

Satellites are also used for weather observation and astronomy.

Remind yourself!

1 Copy and complete:

The ...... a satellite travels through space, the
higher ...... the Earth its orbit is.
Higher ...... take ...... to cover the Earth.
...... satellites have ...... orbits and go around
the Earth often in a 24-hour period.
......-...... satellites take 24 hours to go around
the Earth, circling above the ......

2 How do satellites get the energy they need to
operate?
(Hint: look at the picture at the top
of this page.)

3 Find out about the Hubble telescope.
Why did it work so badly when it was first put into
space?
What did they do to make it work properly?

Think about these questions:

- How big is our solar system?
- How far is it to the nearest star beyond the Sun?
- Why do so many chocolate bars have space-related names?

The first two questions are easy to answer:
The edge of our solar system is about 8 thousand million million metres from the Sun. Our nearest star is Proxima Centauri which is about 40 thousand million million metres away from the Sun. This is a distance of 4.3 **light years**. Light travels at 300 million metres per second.

> A light year is how far light would travel in one year.

Our solar system consists of one star, the Sun and nine planets, most with moons. There is also a belt of asteroids (possibly bits from other destroyed planets) and some comets. When we observe the planets, especially those visible with the naked eye, they look like stars. The planets reflect the light from the Sun but they are not light sources themselves. Reflected light from the planets reaches the Earth, in the same way as light from the Moon does. As they orbit the Sun, the planets move relative to the other stars in the night sky. This makes them easier to detect.

a) Invent a rhyme or a rap to help remember the names of the planets in the solar system.

b) Between which two planets do we find the asteroid belt?

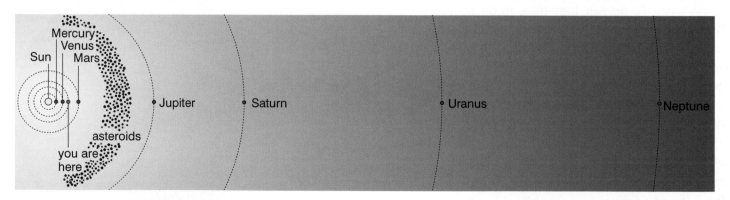

The Milky Way

Our galaxy is called the **Milky Way**. Like many galaxies, it has a spiral arm shape. The arms of the galaxy spin due to the huge gravitational forces that exist. Our solar system is out on one of the spiral arms so we can see most of the Milky Way.

As well as our Sun there are millions of other stars inside the Milky Way. It is likely that many of them also have their own planets orbiting them. We can't see any of these planets through our telescopes because planets do not emit light. Any starlight reflected by these planets would be far too faint for us to observe.

The distance across the Milky Way is approximately 100 000 light years. This is about 100 million million million metres. Quite big really!

Galaxies

Galaxies contain millions of stars. They are spread out over huge distances, millions of times further apart than the planets in our solar system. Some stars are together in pairs called binary star systems. Galaxies also contain the remains of destroyed stars and regions where new stars are being formed.

The Milky Way is one of twenty galaxies that we call the **local group.** Our nearest neighbour is Andromeda. It is only two million light-years away.

Andromeda galaxy.

Constellations

The constellations are patterns of stars that are visible in the night sky. As the Earth moves around its orbit, different constellations become visible at different times of the year. In the southern hemisphere there are other constellations visible. The constellations are formed out of stars that appear to make a shape when they are viewed. People in the past have linked the constellations to different gods. Nowadays those who follow astrology have a keen interest in the position of the Earth compared to the constellations.

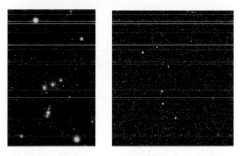

> **c)** Imagine that you had travelled to an Earth-like planet in Andromeda. Would the constellations in the night sky be similar to those on Earth?

The Universe

The Universe contains everything. It includes all the billions of galaxies, each with their thousands of millions of stars. We have a clear idea of how much mass there must be in the Universe. It helps us to estimate the total number of galaxies including any we can't see yet! Scientists believe that the Universe was formed approximately 15 000 million years ago in a huge explosion called **Big Bang**. All the galaxies in the Universe are moving away from the probable site of big bang. We know they are moving because of **red shift**. The light from the stars in distant galaxies is slightly redder than it should be. The galaxies furthest away have the greatest red shift, so they are moving away the fastest. Astronomers have used this information to estimate the age of the Universe.

Remind yourself!

1 Copy and complete:

Our solar system is part of a galaxy called the
.... Galaxies contain ...... of stars which are spread very far apart.
Some groups of stars viewed from the Earth form shapes called ......
The U...... contains billions of galaxies.
It was formed about ...... million years ago by a huge explosion called Big ......

2 If it were possible to build a space ship that could travel at the speed of light. How long would it take to:

i) Get to our nearest star?

ii) Cross the Milky Way?

3 Draw a poster or flow diagram defining the key terms: solar system, Milky Way, galaxies, constellations, Universe.

▶▶▶ 12d Stars

How does it feel to be a star?
Did you know you are made out of **star dust?**
Every atom inside your body was once inside a star!
In fact, apart from the hydrogen, every atom inside
your body was made inside a star.

Stars are massive fusion **nuclear reactors.**
They take atoms of hydrogen and join them together to make helium.
Older larger stars also make atoms of the other elements.

a) How do we know that our solar system was made out of the debris of older generation stars?

Scientists believe that a massive explosion called Big Bang started
the Universe. This is a theory. It is not a proven fact because it all
happened millions of years ago and nobody was there to witness it.
Using the Big Bang to explain the start of the Universe is our best theory, so far!

b) The Earth orbits the Sun. Electrons orbit the nucleus in an atom.
Which of these statements is fact and which is a theory?

The formation of a solar system; a theory

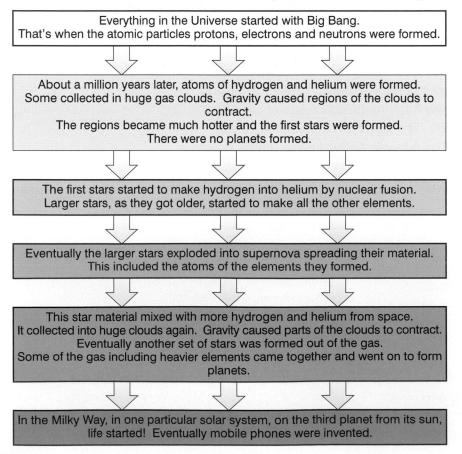

Everything in the Universe started with Big Bang.
That's when the atomic particles protons, electrons and neutrons were formed.

⬇ ⬇ ⬇

About a million years later, atoms of hydrogen and helium were formed.
Some collected in huge gas clouds. Gravity caused regions of the clouds to
contract.
The regions became much hotter and the first stars were formed.
There were no planets formed.

⬇ ⬇ ⬇

The first stars started to make hydrogen into helium by nuclear fusion.
Larger stars, as they got older, started to make all the other elements.

⬇ ⬇ ⬇

Eventually the larger stars exploded into supernova spreading their material.
This included the atoms of the elements they formed.

⬇ ⬇ ⬇

This star material mixed with more hydrogen and helium from space.
It collected into huge clouds again. Gravity caused parts of the clouds to contract.
Eventually another set of stars was formed out of the gas.
Some of the gas including heavier elements came together and went on to form
planets.

⬇ ⬇ ⬇

In the Milky Way, in one particular solar system, on the third planet from its sun,
life started! Eventually mobile phones were invented.

The life of a star

Stars can be put into one of two groups:

A   **1. Smaller mass (yellow star)** ○

This is the stable part
of a star's life (like our Sun).
It uses its hydrogen fuel to make
helium, releasing energy.
A yellow star can continue to generate
heat and light for a billion years.
The heat it generates makes it try to expand
but its gravitational field stops expansion.

B   **1. Larger mass (blue giant) star**

These stars are about
8 times the mass of the
Sun or greater.
This is also the most
stable part of a larger
star's life. Blue giants spend far
less time in this stable stage
than the smaller yellow stars.

2. Red giant

Eventually a yellow star
will use up its hydrogen.
Some nuclear reactions
continue. They use helium
to make oxygen, nitrogen and carbon. The
yellow star expands considerably into a red
giant. It is still hot inside but the outer layer
is cooler.

2. Super red giant

Blue giants reach
the red giant
stage sooner.
They start to make
even larger atoms.

3. White dwarf

Red giants contract
under their own gravity.
The atoms inside it become very densely
packed. The material inside a white dwarf is
about a million times denser than matter on
Earth. A white dwarf will eventually cool and
become a **black dwarf**.

3. Supernova

The super red giant also
contracts but explodes
to become a supernova. A new dense
cloud or a nebula is formed. In the centre
is an extremely dense **neutron star**.
A **black hole** could be formed instead.
Everything near it will be pulled in.

---

c) Imagine you saw a super giant star explode into a supernova last night.
If the star had been 2000 light years away, how long ago did it really explode?

d) When our Sun turns into a red giant, what will happen to life on the Earth?

Remind yourself!

1 Copy and complete:

Smaller mass ...... stars can be stable for a ......
years. Larger ...... giant stars have a ...... life.
Yellow stars eventually expand and turn into ......
giants. These stars contract under their own
gravity and become ...... dwarfs.
Blue giants become ...... red giants. These
eventually ...... forming supernovas, leaving very
dense ...... stars or black holes.

2 Some religions do not accept that the Universe
started with Big Bang.
*Who pushed the button to start Big Bang? How
did we really get here?* Discuss these points in a
small group.
See if you can all agree on a theory.

3 Do some research into supernovas.
Pick one, draw it and then find out its
size and distance from the Earth.

ICT

Why do all aliens on TV speak such good English?
Surely if they were brought up in far distant solar
systems on the other side of the Milky Way, they would
not speak such good English.
TV writers seem certain that life exists outside our solar system.
There seems to be thousands of different species of alien.
Each with a strangely shaped spacecraft which defies every
rule of physics: not only can they move faster than the speed
of light but they can stop instantly without the crew falling over!

We don't really know whether life exists outside Earth.
There are billions of galaxies in the Universe.
Each galaxy has millions of stars.
It's quite likely that some stars would have their own solar systems.
So surely there must be life out there, somewhere … anywhere!

All you need for life (as we know it) is a planet with the right range
of temperatures, not too much gravity, oxygen and some other
useful chemicals. *Or is it?*

a) Between what temperatures can humans live without special protection?

Life in our solar system

Astronomers have been searching for signs of life on the other planets
and moons in the solar system, so far without success. They can look out
for certain chemical compounds that could be made by living things.

b) What gas is produced by plants during the process of photosynthesis?

The amount of oxygen in the Earth's atmosphere is much greater
than it would be without plants producing it.

It is possible that life might have existed on Mars long ago.
There are huge deep valleys on the surface of Mars that people
believe were actually canals. We have already sent some probes
there and received back detailed photographs and other information.
To carry out a more detailed study looking for life, we would have
to travel there. In fact a full expedition is planned within the next decade.
Once there, we could look for fossils of plants and animals.
We could also look for much smaller signs of life such as bacteria
or other microbes.

The surface 'canals' on Mars.

Humans living elsewhere

Ever since we first landed on the Moon, people have dreamt of the idea of a lunar space station. We could build massive domes and set up an Earth-like atmosphere and climate. Water should not be a problem because large amounts of ice have been discovered at the lunar poles. We would have to get used to living with far less gravity though.

Mars also has a lower gravitational field strength than the Earth. Its average temperature is only –20°C and its thin atmosphere contains carbon dioxide. Even so, with a biosphere, water supply and some suitable planting, it could be possible to make Mars habitable.

Europa is one of Jupiter's moons. It is also smaller than the Earth and much further from the Sun. Its surface is thick, very cold ice, but it does have a metal core similar to the Earth. Maybe when the Sun turns into a red giant, humans will be living on Europa instead.

c) How could we check for life forms on Europa? What would we look for?

Meeting extra-terrestrials

It is unlikely that we will ever meet any life forms from other planets. We can't build a space ship that travels faster than the speed of light. Einstein's theory of relativity makes it impossible *(but it's only a theory)*. The same would be true for any aliens wanting to travel here. Physics is physics everywhere. At the fastest speed we could hope to travel, it would take us 4300 years just to get to Proxima Centauri, our nearest star.

d) How fast would we be going if it took us 4300 years to get to Proxima Centauri?

The **search for extra-terrestrial intelligence (SETI)** was set up more than forty years ago. Narrow band radio telescopes are used to listen out for signs of intelligent broadcasts. So far no response, but who knows? We might not be able to get to Proxima Centauri for 4000 years but any extra-terrestrials living on a planet orbiting that star could have been watching our TV sci-fi programmes for the last forty years!

Remind yourself!

1 Copy and complete:

We have not yet found ...... on Mars, the Moon, E...... or elsewhere in our solar ......
To find signs of life involves looking for f......, microbes or chemical ...... from living things.
Using SETI, we have been listening for possible extra-terrestrial broadcasts for 40 years.

2 Design you own biosphere to be transported to Mars.
You need to include in your design brief:

a) Food supplies for a balanced diet.

b) Waste product removal (water).

c) Oxygen and temperature regulation.

d) Energy requirements.

Summary

People used to think that the Earth was the centre of the Universe.
They believed that the Sun and the other planets went around the Earth
and that all the stars were part of a huge roof over everything.

The Earth takes $365\frac{1}{4}$ days (a year) to orbit the Sun.
The Moon is the Earth's **natural satellite**.
It takes just under 28 days to orbit the Earth.
The Earth spins on its own axis once every 24 hours (a day).
Our seasons occur because of the Earth's tilt of 23.5°.

Gravitational attraction between the Sun and the planets makes them orbit the Sun.
The planets orbit in near circular slightly elliptical orbits.
The greater the distance from the planet or star, the lower the pull from its gravity.

The speed of a satellite through space determines its height above the Earth.
Lower orbit satellites travel slower through space than those much higher
above the Earth's surface. Low satellites travel across the surface of the Earth
quickly because their orbits have a small circumference.
Polar satellites are low level satellites that go around the north and south poles.
As the Earth spins beneath them, a different part is visible on every pass.
We often use this type of satellite for military or other monitoring purposes.
We position **geo-stationary satellites** so that they stay above the same
part of the Earth's surface over the equator at all times. They take 24 hours
to complete one orbit. There is only room for about 400 geo-stationary satellites.
They are used for communications, including reflecting TV and radio signals.

Our galaxy is spiral-arm shaped due to gravitational forces. It is called the **Milky Way**.
Galaxies contain millions of stars spread out over huge distances.
The constellations are patterns of stars that we can see in the night sky.
The Universe was formed approximately 15 000 million years ago by **Big Bang**.
All the galaxies in the Universe are moving apart. Astronomers found that the Universe
is expanding by looking for **red shift**; stars moving away from us appear redder.

Stars release energy from nuclear fusion reactions. The main reaction
involves atoms of hydrogen being 'fused' together to form an atom of helium.
Smaller mass yellow stars like our Sun can be stable for a billion years.
Eventually they grow to be red giants and finally contract into very dense
white dwarfs. Blue giants are much larger than our Sun. They remain stable
for less time before expanding into red super giants. These stars explode
forming supernovas with very dense neutron stars or black holes at their centres.

Questions

1 Copy and complete:

i) The Earth is one of the ...... planets that orbit the ...... It takes the Earth ...... days to complete its slightly ...... orbit.
The ...... is the Earth's ...... satellite.
It takes the Moon 28 days to ...... the Earth. The Earth spins once on its axis in ... hours.

ii) G...... attraction makes the planets orbit the Sun. As you move further ...... from a planet or a ......, its gravitational ...... becomes weaker.

iii) Satellites can be positioned in either ...... or high orbits. Low ...... are quicker but the satellites travel through space ...... Polar satellites go around both the ...... and the south pole. ......-stationary satellites are ...... above the equator and they take ...... hours to orbit the Earth. They are always above the same places so they can be used for reflecting TV and ...... signals.

iv) Our galaxy is called the ...... Way.
There are ...... of galaxies, each containing millions of ...... Together they make up the ...... The Universe is ex......, which is shown by red shift.
The stars appear ...... than they really are so they are moving ...... from us.

2 i) What would happen to the position of winter if there were no leap years? (February the 29th only happens every four years.)

ii) Why doesn't light reflected off the Moon reach the Earth during a new moon?

iii) What would the UK's climate be like if the Earth was not tilted?

iv) How would having no tilt on the Earth affect the length of the day?

v) How do you think it would affect human sleep patterns if the Earth took 48 hours to rotate on its axis?

3 i) Use a spreadsheet or graph paper to draw a bar chart showing this planet data.

Planet	Distance from Sun (1000 million metres)	Time to orbit the Sun (Years)	Surface gravity (N/kg)
Mercury	60	0.25	4
Venus	110	0.6	9
Earth	150	1	10
Mars	230	2	4
Jupiter	780	12	25

Use your graph to answer these questions.

ii) What is the pattern between the distance from the Sun and the time to orbit once?

iii) Is there a relationship between the gravity values and the other information?

iv) Which planet has the closest gravity to the Earth? Could we live there?

v) Which planet has the most gravity?

4 i) Which is further away from the Earth, the Moon or a geo-stationary satellite?

ii) Which travels fastest through space, a polar satellite or the Moon?

iii) Find out when the first satellite was launched into space.

5 Use information from page 159 to construct a flow diagram showing the life cycles of a yellow, lower mass star and a blue giant star.

6 Imagine that you were able to travel out of our solar system in search of a planet to live on. (You had a special space ship that could travel much faster than the speed of light.)
Make a list of the different properties that your chosen world would need to sustain human and plant life.

1 This question is about machines in a fitness centre.

(a) The diagrams show a student using a rowing machine and its display panel.
The display panel shows the readings at the end of the exercise.

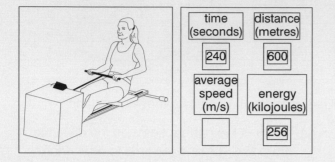

time (seconds)	distance (metres)
240	600
average speed (m/s)	energy (kilojoules)
	256

Calculate the missing reading on the display panel. (2)

(b) The diagrams show a student using a step machine and its display panel. The display panel shows the readings at the end of the exercise.

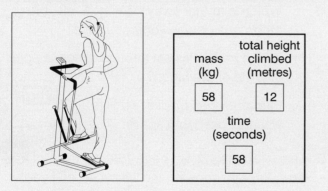

mass (kg)	total height climbed (metres)
58	12
time (seconds)	
58	

The equation below is used to calculate weight.

weight = mass × gravitational field
(newton, (kilogram, strength
N) kg) (newton/kilogram, N/kg)

Use this equation to help you calculate the work done by the student during the exercise. (4)

(AQA 2001)

2 A steel ball was dropped and fell 180 m.
The graph shows how far it had fallen at different times.

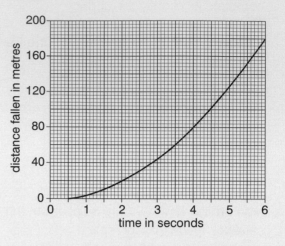

(a) How far had the ball fallen after 4 seconds? (2)

(b) Explain how the graph tells you what happened to the speed of the ball as it fell. (2)

(EDEXCEL 2000)

3 (a) The graph represents a cycle journey from home to school.

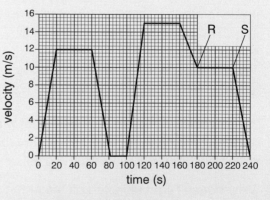

On a copy of the graph:

(i) mark with the letter **X** a part where the cyclist was moving with constant velocity;

(ii) mark with the letter **Y** a part where the cyclist was moving with constant acceleration. (2)

(b) Calculate how far the cyclist travelled between points **R** and **S** on the journey. Show your working giving your answer in metres. (3)

(AQA (NEAB) 2000)

4 A table tennis ball is dropped on Earth.
Another table tennis ball is dropped on the Moon.
The graph shows what happens to each ball
during the first five seconds of its fall.

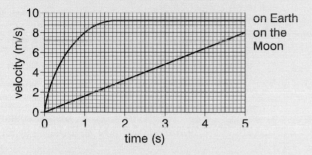

(a) On the Moon, the ball takes 5 seconds to reach
a speed of 8 metres per second. Calculate the
acceleration of the ball. (Show your working
and give the unit). (3)

(b) Describe the differences between the
acceleration of a ball on Earth and the
acceleration of a ball on the Moon during the
first five seconds of their fall.
(You do not need to do any calculations.) (3)

(c) Suggest why the acceleration of a ball on the
Moon does **not** change. (1)

(AQA 2001)

5 A sprinter can run 200 m in 25 s.

(i) Write down, **in words**, an equation connecting
speed, distance and time. (1)

(ii) Calculate the speed of the sprinter in m/s. (2)

(WJEC 1999)

6 The stopping distance of a car increases when the
road is wet.
The diagrams show the tyre of a car driven at two
different speeds on a very wet road.

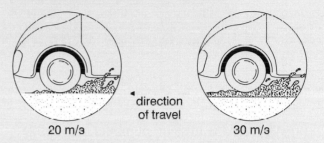

(a) Driving at 30 m/s on a very wet road can be
dangerous.
Use the diagrams to help you to explain why.
(3)

(b) A car is moving along a dry road.
The driver sees danger ahead. It takes him
0.7 s to react before braking. This is called the
reaction time.
In this time, the car travels 21 m.

(i) Calculate the speed of the car during this
0.7 s. (2)

(ii) The distance travelled before braking is the
thinking distance.
The stopping distance of the car is:

thinking distance + braking distance

The stopping distance at this speed is
93 m.
What is the braking distance of the car? (1)

(iii) What effect does driving on a wet road
have on the driver's reaction time? (1)

(EDEXCEL 1999)

7 The table gives data for the overall stopping distance of a car at different speeds.

Speed of car (m/s)	Thinking distance (m)	Braking distance (m)	Overall stopping distance (m)
5	3.5	2.0	5.5
10	7.0	8.0	B
15	A	18.0	28.5
20	14.0	32.0	46.0

(a) What are the missing numbers in the table labelled A and B? (2)

(b) The **thinking** and **braking** distances depend on the conditions in which the car is being driven.

(i) Which **two** factors below would **increase** the thinking distance.
the driver drinking alcohol
the car moving faster
good visibility
dry roads
worn brakes (2)

(ii) Which **two** factors below would **increase** the braking distance.
the driver drinking alcohol
the car moving faster
good visibility
dry roads
worn brakes (2)
(AQA 2001)

8 (a) Rewrite the descriptions below and then match them to their names.

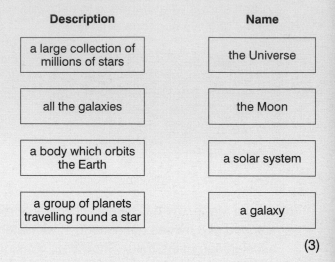

Description	Name
a large collection of millions of stars	the Universe
all the galaxies	the Moon
a body which orbits the Earth	a solar system
a group of planets travelling round a star	a galaxy

(3)

(b) Why do planets remain in orbit round the Sun, and not move off into space? (2)
(EDEXCEL 1998)

9

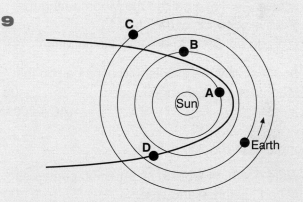

The diagram shows the orbits of some bodies around the Sun. The arrow shows the direction of the Earth's orbit.

(a) Choosing from **A**, **B**, **C** and **D**, state which body is

(i) a comet (1)

(ii) Venus (1)

(b) Mark on a copy of the orbit of **A** an arrow to show the direction in which it moves. (1)

(c) (i) Describe the shape of the orbit of **D**. (1)

(ii) Name the force which keeps **D** in its orbit. (1)
(WJEC 1998)

Section Four
Waves

In this section you will see how waves transfer energy.
You will learn about the way waves travel through substances
and what happens when they change speed or direction.
You will find out that some waves can be used for communication
while others can give us cancer.

CHAPTER 13

WAVE PROPERTIES

▶▶▶ 13a Types of wave

When asked about waves, most people think of the sea.
Many leisure pools have wave machines. It's great fun
being swept along by the force of the water.

> Waves transfer energy from one place to another.

Think about the rough seas that you see in films or on the news.
Imagine the amount of energy those waves transfer!

a) What damage could large waves cause when they hit the coast?

Water waves are not the only type of wave.
Shake one end of a rope while the other end is held.
You will make a wave pass along the rope.
This is exactly what you are doing when you pluck a guitar string.

Vibrating the guitar string creates a sound wave,
which travels through the air to your ear.
Anything vibrating can be the source of a sound wave.

b) What type of wave does a vibrating guitar string produce?

The sound waves hitting objects will cause them to vibrate.
This transfers energy into the objects. The speaker's sound wave
will make your ear drum vibrate so you hear the sound.
As you move away from the speaker, the sound waves spread out,
so the energy they transfer is reduced. The sounds become less **intense**.
As long as the volume of the speaker is kept the same, the sound wave
will be quieter the further away from the source that you move.

> The amount of **energy** that a wave is able to **transfer** reduces
> as the wave spreads out, moving away from its source.

c) As you move away from a speaker, what happens to the loudness
of the sound wave?

Wave travel

Most waves need a material or **medium** to travel through.
Some waves make the particles of the medium move up and down.
Other types of wave make the particles move backwards and forwards.
Either way:

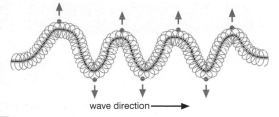

wave direction ⟶

The **particles** of the medium **do not get carried** along with the wave.

The wave passes on leaving the particles of the medium behind.

d) What happens to a boat when a water wave passes underneath it?

Transverse waves make the particles of the medium move at right angles to the direction that the wave is travelling.
The particles vibrate **perpendicularly** to the direction of movement.
When a Mexican wave passes along the audience in an all seater stadium, none of the people move with the wave. They only move their arms up and down to help the wave to continue through the stand.
Mexican waves, water waves and the waves passing through a rope or a stringed musical instrument are all examples of transverse waves.
Light and radio waves are examples of a group of very important waves called **electromagnetic radiation**. These are also transverse waves even though they can travel with no material to pass through.
They can travel through the vacuum of space (see Chapter 15).

Longitudinal or compression waves cause the particles of the medium to vibrate backwards and forwards.
Sound waves travel as longitudinal waves. They can travel through solids, liquids and gases. When a sound wave travels through a material, the particles inside it are compressed closer together and then pulled further apart. Another example of a longitudinal wave is a seismic P wave (see page 181). Longitudinal waves need to have a medium to compress and stretch, so they can't pass through a vacuum.

wave direction ⟶

Remind yourself!

1 Copy and complete:

Waves ...... energy from a source to its surroundings.
A water wave is an example of a ...... wave.
Water particles are caused to vibrate ...... and down. They move at ...... angles to the direction of the wave. All electromagnetic waves including ...... travel as ...... waves.
Longitudinal ...... make the particles of the medium they travel through ...... backwards and ...... in the same direction as the wave. Sound travels as a ...... wave.

2 In each of the following, which transfers more energy:

i) A small water wave or a tall one hitting a sea wall?

ii) A 150 W lamp or a 12 W night lamp viewed from one metre away.

iii) An explosion 1 mile away or the headphones from a loud personal CD on full volume.

3 Why do you think sound travels through wood much better than wool?
They are both solids.

You might have noticed that boys have deeper voices than girls.
Does this mean that boys talk louder than girls?

Boys' voices are mostly a lower pitch than girls'.
Pitch is not the same as loudness so the boys do not automatically
talk louder (it just seems that way).
The pitch of a sound wave depends on its frequency. A loud sound
can be high pitched (high frequency) such as a smoke alarm,
or low pitched (low frequency) like a road drill.

Frequency describes how many times something happens in a second.
While you are exercising, your heart might pump 120 times every minute.
Your pulse rate is 120, which is 2 beats per second.
The frequency of your heart pumping is 2 hertz (Hz).
Frequency is a very important term when describing a wave.

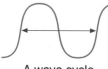

A wave cycle
(wavelength)

> The frequency of a wave is the number of complete **wave cycles** per second.

The waves might be coming from a source or just passing a point.

> **a)** A group of pupils running down a flight of stairs.
> Five pupils pass the bottom stair every second.
> What is the frequency of their escape?
>
> **b)** A sound wave from a speaker has a frequency of 20 Hz.
> How many waves leave the speaker every minute?

What is meant by a 'complete cycle' of a wave?
Picture a water wave. Each high crest is followed by a low
trough and then another high crest again. One complete cycle
is the distance from one crest to the next crest.
We call this distance the **wavelength** (symbol, λ).

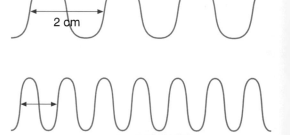

2 cm

> Wavelength (λ) is the length of one complete cycle of a wave.

Wavelength can be measured from trough to trough,
or from midpoint to midpoint. When wave crests get closer
together, the wavelength is shorter.

> **c)** What are the wavelengths of the waves shown on the right?

Unlike a water wave, you can't see a sound wave.
The wavelength of a sound wave is the distance from
one compression point to the next compression point.

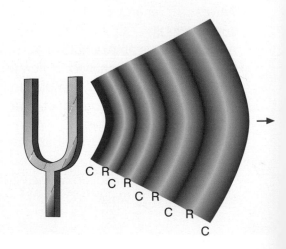

C R
 C R
  C R
   C R
    C

> **d)** Sketch and label another sound wave with twice the
> wavelength of the wave shown on the right.

The speed of a wave

A wave's speed through a medium
depends directly upon its frequency and its wavelength.

Equation 14 to learn

	wave speed	**=**	**frequency**	**×**	**wavelength (λ)**
Symbol form	v	$=$	f	$\times$	λ
	metre/second (m/s)		hertz (Hz)		metres (m)

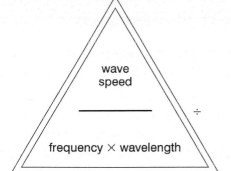

Example A water wave travelling across a leisure pool.

Frequency of wave = 2 Hz
Wavelength of wave (λ) = 1.5 m
Using wave speed = frequency × wavelength
$$= 2 \text{ Hz} \times 1.5$$
$$= 3 \text{ m/s}$$

e) What is the speed of a rope wave? Its frequency = 3 Hz, its wavelength λ = 4 m.

The disturbance a wave causes (its size/height) is called its **amplitude**.

> Waves with more energy have a larger amplitude.

A large water wave has high crests above and deep troughs
below a flat sea. Its large amplitude disturbs or disrupts
the flat sea (up and down).
Loud sound waves have large amplitudes, quiet sounds have
small amplitudes.

f) What is the amplitude of the two waves on the right?

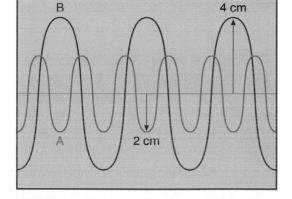

Remind yourself!

1 Copy and complete:

The ...... of a wave is the number of waves per
......

The ...... (symbol ...) of a wave is the length of
one complete ...... of the wave.

   Wave ...... = frequency × ......
Symbol form v = ...   × λ

The disturbance caused by a wave passing is
called the ...... Loud sounds have a ......
amplitude; they transfer more energy.

2 Sketch the following waves:

i)   A wave with an amplitude of 1 cm and
     wavelength of 3 cm.

ii) A wave with an amplitude of 2 cm and a
     wavelength of 1 cm.

3 Calculate the following:

i)   Wave speed of a water wave with a frequency
     of 0.25 Hz, wavelength λ = 8 m.

ii) The frequency of a sound wave from a guitar
     string, speed 340 m/s, wavelength 0.5 m.

Sometimes shop windows make good mirrors if you're outside. Imagine how amused the people inside the shop are at the strange faces you pull at the window while you check your hair.

When you look in a mirror you see your reflection.
Some of the light hitting your face will travel on to hit the mirror.
It will be reflected back into your eyes and you will see your **image**.
When you use a shop window as a mirror, some light is being reflected back to your eyes from the glass surface. Most of the light from your face will pass straight through the glass into the shop.

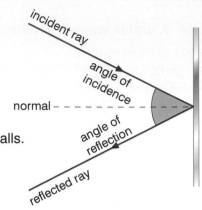

Reflection

Any type of wave can be reflected. A water wave on the surface of a swimming pool will reflect or bounce back off the smooth sides of the walls. It will reflect just like a light ray off a plain mirror, or sound wave off a wall. The angle of incidence for a wave is the angle away from the normal that the wave hits the surface (normal = 90° to surface). Waves reflecting off plain smooth surfaces follow the rule of reflection:

    Angle of incidence = the angle of reflection.

Work out the angles of reflection for these rays of light hitting a plain mirror:
a) Incident angle = 30°, **b)** Incident angle = 45°, **c)** Incident angle = 0°

A wave hitting a surface at 90° (travelling down the normal) will be reflected straight back the way it came. This is what happens when you look straight at a mirror or when a sound wave rebounds off a wall and you hear your echo.

When a ray of light hits a rough or matt surface, it is spread out in all directions and no reflection is seen.
Any type of waves will do the same at a rough surface.
Each part of the wave will be reflected off at a slightly different angle.

    Rough surfaces disperse waves.

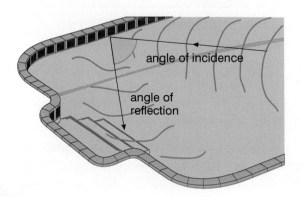

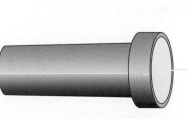

Refraction

Glass is a **transparent** material. It wouldn't be much use for windows if it wasn't. Perspex and water are also **transparent**. They allow light to pass through them, but does the light change?

| The speed of a wave depends on the medium it is travelling through. |

Light travels fastest in a vacuum. It travels much faster in air than it does in glass, perspex or water. Light entering a glass block from air will slow down. When it goes back into the air, it speeds up again.

Which is easier to run across, a sandy beach or one covered with 20 cm of water?

Light hitting a transparent surface perpendicularly (at 90°) will pass straight through without altering its path (Light ray A).

| Waves crossing a boundary at 90° will change speed but not direction. |

If a wave **crosses a boundary** between two mediums **at an angle** to the normal it will change its direction as well as its speed. This is called **refraction**. A ray of light (B) entering a glass block is refracted (*bent*) in towards the normal because it slows down.

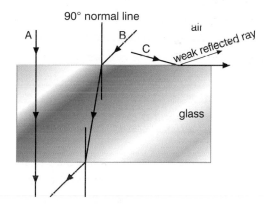

d) How will a ray of light be refracted when it leaves a glass block?

Some light is always reflected back at the boundary, this is why you can see yourself in the shop window. If the incident angle of the light hitting the block is increased, eventually the light stops entering (C). It gets reflected off the surface of the block, just as if it were a mirror.

When you swim underwater, the sounds from above are completely different. The sound waves have been refracted as they enter the water. A water wave can also refract when it passes from deep to shallow water if it hits the depth change at an angle. Look at the picture on the right.

Remind yourself!

1 Copy and complete:

All ...... can be reflected or refracted.
A wave will be reflected if it hits a plain ......
surface. If the surface is rough, the wave will be
......
Refraction occurs when a wave ...... a boundary
from one medium to another at an ...... Light
slows down and is refracted ...... the normal
when it enters glass from air.

2 Students find it difficult not to get confused by the two 'R' words, reflection and refraction. Invent a way of remembering:

i) The law of reflection at a plain surface.

ii) What happens when light is refracted when it enters a glass block.

3 When a ray of white light is directed through a prism, it is split up into the colours of the rainbow. Violet is refracted the most. Which colour light is refracted the least?

▶▶▶ 13d Diffraction

Imagine you are sitting in your classroom working on an important science test. It's a hot summer day so the doors are open. Suddenly a bunch of year seven students come running down the corridor. You can hear them but you can't see them until they pass the door.

You can't see the noisy students because light travels in straight lines. It can't bend around the door to travel to your part of the classroom. You can hear them because the sound waves they are producing are able to spread out into your classroom so they can pass to your ear.

The same thing happens when you are hiding behind a wall, trying to avoid cross country. You listen to check that nobody has spotted you. Any sound they make will be spread out by the top of the wall so you will hear it. To be certain you are safe involves looking over the wall. That's when you get spotted and are given the detention.

a) If light could spread around corners would you still get shadows?

> **Diffraction** is a property of waves.
> It is the spreading out of waves after they have passed through a gap.

The gap needs to be a similar size to the wavelength of the wave. Sound diffracts easily because the wavelengths of audible sound waves range from approximately 1 cm to 20 m. Most door openings are more than 1 cm but far less than 20 m. This means that a sound wave will spread out after passing through a doorway. Sound being diffracted gives us evidence that sound travels as a wave.

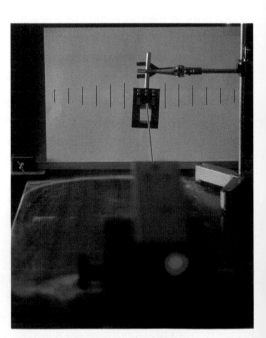

Visible light has wavelengths around 0.000 000 1 m. This wavelength is far smaller than any doorway, so light will not spread out as it passes.

b) The wavelength of light is very small. Do you think it is possible to get a small enough gap to pass light through to make it diffract?

Light can be diffracted which is one of the reasons that we know that light travels as a wave. The picture on the right shows a laser beam being sent through a diffraction grating. The laser light is diffracted making it form the pattern shown. A diffraction grating is a bit like thousands of tiny doorways which make the single laser beam spread out.

A laser beam being diffracted.

Diffraction of other types of wave

Look at the ripple tank pictures on the right.
The first pattern shows the diffraction of a wave leaving a gap.
The gap is close in size to the water wave's own wavelength.
Light leaving a diffraction grating or sound coming through
a door would have a similar pattern.

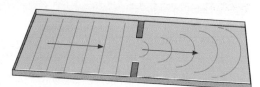

Simple diffraction pattern.

The second picture shows how diffraction happens at the edge
of a wave when it passes close to an object. As the wave moves
past the object, its edge starts to spread out or diffract.

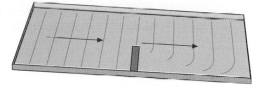

c) Use this example of diffraction to explain how you can
still hear people in the garden next door from behind a tall wall.

*Diffraction of the edge of a wave having
just passed an object on one side.*

Water waves are much easier to diffract than light beams because
they have much longer wavelengths.

The longer the wavelength of a wave, the more easily it will be diffracted.

Have you ever had trouble picking up a signal on your mobile phone?
Mobile phones use microwaves. Microwaves, radio waves and visible
light are all part of the electromagnetic spectrum (see Chapter 15).
Long wave radio can have wavelengths of over 1000 m.
Microwaves have wavelengths of a few centimetres.

d) Which is more likely to diffract around a building or behind a hill,
a microwave or a long wave radio transmission?

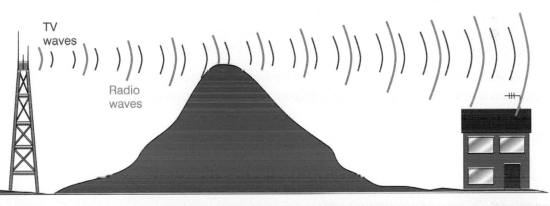

TV
waves

Radio
waves

Remind yourself!

1 Copy and complete:

Diffraction is the ...... out of a wave when it
moves through a small ...... or past an obstacle.
All ...... can be diffracted.
The gap a wave passes through needs to be a
similar size to its ......
The ...... the wavelength of the wave, the more it
will diffract.

2 a) Look at the first ripple tank diffraction patterns
above. What would happen to the pattern if:

i) The gap was made smaller?

ii) The water's wavelength was increased?

b) Sketch the pattern you would see if the gap
was made much larger.

Summary

Waves **transfer energy** from one place to another.
The amount of energy that a wave transfers to its surroundings
reduces as the wave spreads out, moving away from its **source**.

Most waves need a material or **medium** to travel through.
The particles of the medium do **not** get carried along with the wave.
Transverse waves make the particles of the medium they travel
through move at right angles (up and down) to the wave's direction.
Examples of transverse waves include surface water waves, waves
in ropes and all forms of **electromagnetic radiation**.
Longitudinal or compression waves cause the particles of the medium
they travel through to vibrate backwards and forwards.
They can travel through solids, liquids and gases.
Sound waves and **seismic P** waves travel as longitudinal waves.

The **frequency** of a wave is the number of complete wave cycles per second.
The waves might be coming from a source or just passing a point.
Wavelength (λ) is the length of one complete cycle of a wave.
Wavelength is normally measured crest to crest.
When the crests get closer together, the wavelength is shorter.

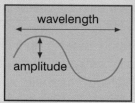

Equation 14

$$\text{wave speed} = \text{frequency} \times \text{wavelength } (\lambda)$$
metre/second (m/s)   hertz (Hz)   metres (m)

The disturbance a wave causes is called its **amplitude**.
Waves with more energy have a larger amplitude.

All waves can be **reflected**: the angle of incidence = the angle of reflection.
Rough surfaces disperse waves, reflecting them in all directions.

All waves can be **refracted**: a wave crossing a boundary between two
mediums at an angle to the normal will change its direction and speed.
If the wave slows down, it is bent towards the normal.

All waves can be **diffracted**: they spread out after passing through gaps.
The gap needs to be a similar size to the wavelength of the wave.
The longer the wavelength of a wave, the more easily it will be diffracted.

As light and sound can be reflected, refracted and diffracted,
it helps to confirm that they are types of wave.

Questions

1 Copy and complete:

i) Waves transfer ...... from a source.
As waves ...... out, the amount of energy they can pass to the surroundings ......

ii) Most waves cause the particles in a ...... they pass through to vibrate.
The particles do not ...... with the wave; their vibrations just help the ...... move.

iii) Transverse waves make ...... move up and ...... Surface ...... waves and the waves in ropes or string are ...... waves.

iv) Longitudinal waves make particles vibrate ...... and forwards. They can travel through gases, ... and solids.
Sound travels as a ...... wave.

v) At a smooth plain surface, the angle of ...... is the same as the ...... of reflection.

vi) Waves get ...... when they move from one ...... into a different medium.
Their speed and direction will ...... when they cross the boundary at an ......

vii) Waves will be ...... when they pass around an ...... or through a ......
The amount they are diffracted depends on their w...... The closer it is in size to the ...... of the gap, the more the wave will get ...... Waves with ...... wavelengths are diffracted the most.

2 Sketch a slinky spring carrying:

i) A transverse wave.
Label the wavelength and the amplitude.

ii) A longitudinal wave.
Label the regions of compression and rarefaction (stretched out).
Label the wavelength.

3 a) Draw the following transverse waves:
i) Wavelength 4 cm, amplitude 2 cm.
ii) Wavelength 2 cm, amplitude 4 cm.

b) If both waves were travelling at the same speed, which would have the highest frequency?

4 A group of students using a rope create a number of different transverse waves. They change the length of rope and the frequency of making waves.
(They measure the frequency of their waves by timing ten complete waves and then dividing.)
Copy and complete this table of their results using the wave speed equation.

Frequency (Hz)	Wavelength (m)	Wave speed (m/s)
0.5	4	
1	0.5	
2		2
	4	1

5 A sea wave travels 180 m in 60 sec.

i) Use equation 12 from page 130 to calculate the water wave's speed.

ii) The sea wave's wavelength is 6 m.
Use the wave speed equation to calculate its frequency.

6 At room temperature all sound waves travel at approximately 340 m/s.

a) Calculate the frequencies of the following sound waves:
i) Wavelength = 0.5 m.
ii) Wavelength = 1 m.
iii) Wavelength = 2 m.

b) What do you notice about the frequency values as the wavelength is doubled?

7 Look back at the sections on reflection, refraction and radiation.
List as many points as you can find that show either light or sound behaving like a wave.

CHAPTER 14

SOUND, ULTRASOUND and DIGITAL SIGNALS

▶▶▶ 14a Sound

How often has a teacher made you work in silence because of 'too much noise'?

It's virtually impossible to get the silence the teacher wants.

You always get: Chairs and stools dragging across the floor;

Coughing that seems to spread around the room;

Noise from other classrooms echoing everywhere;

Worst of all, the clock won't stop ticking.

It seems that the quieter you try to sit, the more noise there is.

Some of the properties of sound were discussed on pages 168–169.
Sound waves:

- are caused by vibrations;
- transfer energy;
- spread out from their source;
- travel as **longitudinal** waves;
- need a medium (solid, liquid or gas) to travel through;
- make objects vibrate when they hit them.

a) Put the information about sound waves given above into a simple table.

b) What do the terms frequency and amplitude mean? (See pages 170–171.)

Frequency and sound waves

The frequency of a sound wave is the number of vibrations that its source makes per second. Pitch is directly related to frequency: a drum produces a low pitched (low frequency note) and a whistle gives a higher pitched (high frequency) note.

c) What is the lowest pitched sound you can hear?

d) How do you feel when you hear a very high pitched sound?

The limits of humans' hearing ranges from about 20 Hz to around 20 000 Hz (20 000 Hz is 20 kHz).

e) Name an animal that can hear sounds above 20 kHz?

f) When you talk loudly instead of quietly, do you think that you produce sound waves of a higher frequency?

Amplitude and sound waves

The amplitude of a sound wave is its loudness.
The louder a sound, the more energy it transfers to objects it hits.

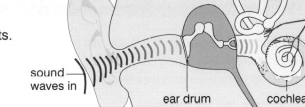
sound waves in / ear drum / cochlea

> The unit of loudness is the **decibel** (dB).

A whisper is around 20 dB and a road drill nearby is about 100 dB.

Loud sounds can damage your hearing. They make the fluid inside
the cochlea in your inner ear move violently. As the fluid vibrates
it can damage the tiny hairs that help you hear particular frequencies.
The hairs connect to the auditory nerve. Longer hairs pass
on low pitched sounds; shorter hairs pass on high pitch.
This is why overexposure to loud sounds can lead to some deafness.

> g) Which cochlea hairs do you think get lost first in people who
> spend too much time exposed to loud music in clubs?

Representing sound waves on an oscilloscope

You might have seen a cathode ray oscilloscope (CRO) used to show
pictures of sound waves. The waves appear as transverse
waves on the display. This is because the microphone
picks up how the sound wave affects air pressure.
The wave crests are the air **compressed above normal air pressure**.
The wave troughs are the air **rarefaction regions**, below normal air pressure.

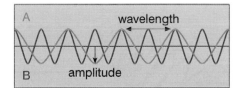

Trace 1. The trace shows two waves
with the same amplitude. Wave A has
half the frequency of wave B.

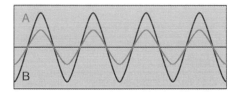

Trace 2. The trace shows two waves
with the same frequency. Wave A has
half the amplitude of wave B.

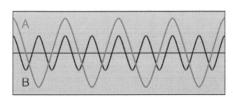

Trace 3. The trace shows two different
waves. Wave A has half the frequency
but twice the amplitude of wave B.

Remind yourself!

1 Copy and complete:

Sound ...... are caused by vibrations. They are
...... waves and they need a medium to ......
through.
The ...... of a sound wave relates to its ......
High pitch is a ...... frequency. Humans' hearing
range is 20 Hz to ...... Hz.
The amplitude of a sound wave is its ......
The unit of loudness is the ...... (dB).

2 Look carefully at the wave trace diagrams from a
CRO above.
Sketch the following pair of sound waves.

Sound wave A, amplitude = 2 cm
            wavelength = 2 cm
Sound wave B, amplitude = 2 cm
            wavelength = 1 cm

If both sound waves are travelling at the same
speed, which has the higher frequency?

Look at the photograph of a baby still inside its mother's womb.
It wasn't taken with a normal camera,
so how do they take pictures like this?

The photo is from a pre-natal scan.
It is a print out of the image formed during an **ultrasound scan**.

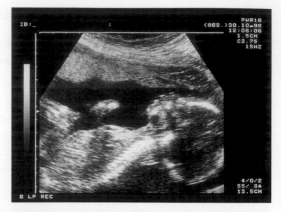

Ultrasound scan of fetus.

Ultrasound

Humans can't hear sounds above 20 kHz.

> Ultrasound is high frequency sound beyond our range of hearing.

Dogs hear frequencies up to about 40 kHz.

a) Why can't humans hear some high pitched dog whistles?

Bats use ultrasound to fly. They send out sounds and listen for
any echoes from nearby objects. The echoes help them to gain
an ultrasonic picture of their surroundings. This is called **echo sounding**.
Ship sonar systems use echo sounding with ultrasound just like bats.
They constantly monitor the depth of the sea beneath them.

b) Describe how a bat uses sound waves when flying, to avoid objects.

Speakers convert electronic signals into sound waves (see page 95).
Ultrasound transmitters are like special speakers just used to produce
high frequency **ultrasonic** waves (ultrasound).
Ultrasound machines send out thousands of tiny bursts of very
high frequency ultrasound (1 to 2 MHz). In between bursts,
they wait to collect any ultrasound that is reflected back.
The ultrasound machine then measures the time and direction of the
ultrasonic echoes to create three dimensional images.

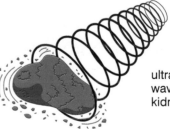

ultrasound
wave hitting
kidney stone

In medicine, ultrasound machines are used to image many other parts
of the body as well as for pre-natal scanning. Ultrasound can be used
to detect and then destroy kidney stones by giving them tiny shock waves.
Broken bones which have an ultrasonic wave passed through them
can be encouraged to heal quicker. Ultrasound scans can also reveal
cancerous tumours.

In industry, ultrasound is used to dislodge tiny bits of dirt in awkward places
within fine machinery or jewellery. The ultrasound *shakes* the dirt loose
without damaging anything, providing a safer method of cleaning the dirt away.

c) Give four uses for ultrasonic waves.

Echo sounding and the Earth

Many house walls are thin layers of plaster board attached to wooden beams. If you ever have to attach a screw to that type of wall, remember **echo sounding**. You can't use ultrasound like bats or a ship's sonar, but you can tap the wall. As you move your hand along it, you will detect the wooden beams between the hollow sections. So you can avoid fixing things into weak plaster. But be careful; heating pipes and electricity cables don't sound hollow either!

Geologists use exactly the same idea to study rock structure in the search for gas and oil. They send ultrasonic waves down into rock layers. If they bounce off the right type of hard rock, there is a chance that oil might be found further down. Once possible sites have been found, the next task is to drill!

Seismic waves

Seismic waves are produced by earthquakes. They travel through the Earth so they have been very helpful in assisting geologists to work out the structure of the Earth.

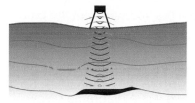

> Seismic **P waves** are longitudinal waves (compression waves).
> Seismic **S waves** are transverse waves, often called **s**hear waves.

Only P waves can travel through both liquid and solid rock. The S waves cause liquid rock particles to just slide over each other until the wave's energy is completely spread out.

earthquake

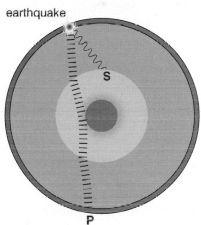

> **d)** How do P waves make the rock particles vibrate: up and down or backwards and forwards?

Geologists use **seismographs** to study the strength and location of earthquakes. Geologists discovered that only P waves would show up on seismographs from an earthquake on the opposite side of the Earth. The S waves did not make it across, stopping at the core. They decided that part of the Earth's core is liquid (the outer core).

Remind yourself!

1 Copy and complete:

Ultrasonic ...... (ultrasound) are high ...... sound waves above the ...... limit of hearing.
Pre-...... scans use ultrasound to make an image of a baby still in the womb.
Echo ...... by reflecting ...... is also used by bats, in ships ...... and to study the Earth.
Seismic ...... given out in ...... quakes are also used to study the structure of the ......

2 A ship sonar operator knows the speed of ultrasonic waves in water. If they timed how long it took a sonar burst to return, how could they find the depth of the sea? (Hint: equation 12, speed.)

3 Do some research into earthquakes. Find out which type of wave travels fastest, and which can do the most damage?

What type of watch have you got, analogue or digital?
Does it have a clock face with hands (analogue)
or a number display (digital)?
Nowadays many watches have both types of display,
but it is not that long ago that they were all analogue.
They didn't have batteries either, you had to wind them up every day!

Have you ever used a voltmeter to measure voltage?
Some voltmeters have number displays that show the voltage
value instantly. The other type of meter uses a needle and a dial.
You have to work out the voltage value by seeing where
the needle points on the dial.

a) Which of the voltmeters in the picture has an analogue display?

Different types of signal

Look at the two wave diagrams below.

b) Which wave diagram do you think represents a digital signal?

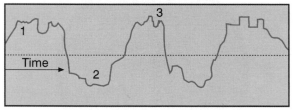

Wave A. A typical sound wave showing the amplitude
and frequency changes as someone talks.

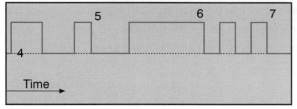

Wave B. A trace of a signal being processed through
a micro chip inside a computer.

Wave A represents an analogue signal
The values of the amplitude and the frequency change as the wave travels.
At point **1** the amplitude is positive while at point **2** it is negative.
The wave crest at point **3** has its own set of little crests and troughs
because the frequency is changing.

> The **amplitude** of an **analogue** signal **continually changes**.
> Signals that have a range of frequencies are also analogue signals.

Wave B represents a digital signal
The amplitude values for the wave signal B have two values: **on** or **off**.
Values on the axis at **4** are off, values level with **5** are on.

> Digital signals are pulses with only two possible values, on or off.

c) Which point shows a longer **on** pulse on wave B, 6 or 7?

Digital communications

Over the last 20 years there has been a communications revolution.
Advances in computer technology have had a major effect
on every part of our lives.

d) How old were you when you first used a computer?

Computers process numbers, that's all they really do.
It doesn't matter how fancy the software, how colourful the graphics
or how clear the sound from CDs are, all computers can ever do is count.
The deepest part of any computer program uses binary numbers (0 and 1).
They are processed as digital signals, **on**'s and **off**'s.

The first popular computer used in schools.

Telephone systems transfer many forms of modern communication.
We use them to talk to friends, send faxes, connect to the Internet,
and transfer text messages between mobile phones.
The amount of information that telephone systems have to
carry has increased dramatically with the communications revolution.

e) Estimate how many phone calls you make and the number
of text messages you send and receive per week.

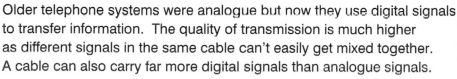

TV with digital decoder.

Older telephone systems were analogue but now they use digital signals
to transfer information. The quality of transmission is much higher
as different signals in the same cable can't easily get mixed together.
A cable can also carry far more digital signals than analogue signals.

We can also now use optical fibres instead of wire cables.
An optical fibre can transfer far more digital signals
than a wire cable of the same diameter.
Optical fibres transfer information as visible light or infrared rays.
They can't carry electrical signals so these must be converted first.
Optical fibres are discussed further on page 191.

Optical fibres.

'Traditional' telephone cable.

Remind yourself!

1 Copy and complete:

Waves can transfer information as either …… or
as digital ……
A digital signal has two states, …… or off.
Computers process digital ……
Communication transfers that use …… signals
are a higher …… than analogue signals.
Digital electrical signals are converted into ……
or infrared rays and sent through …… fibres.

2 Music cassettes and video tapes record sound
and TV signals as analogue signals. How are
music and video signals recorded on mini discs,
CDs and DVDs?

3 What is the average number of digital TV
channels or digital radio stations used per
person for the students in your class?
Carry out a simple survey to find out.

Summary

Sound waves are caused by vibrations and travel as **longitudinal** waves.
They need a medium (solid, liquid or gas) to travel through.

The **frequency** of a sound wave is the number of vibrations per second.
Pitch is related to frequency: drums give low pitched, low frequency notes.
Humans can hear sounds of frequencies from 20 Hz to around 20 000 Hz.
1000 Hz is 1 kHz, a million Hz is 1 MHz.

The **amplitude** of a sound wave is its loudness.
The louder a sound, the more energy it transfers to objects it hits.
The unit of loudness is the **decibel** (dB).
A whisper is around 20 dB and a road drill nearby is about 100 dB.

Sound waves can be shown on a **cathode ray oscilloscope**.
They appear as **transverse waves** on the display.
The wave crests show the air **compressed** above normal air pressure.
The wave troughs show **rarefaction** (air below normal air pressure.)

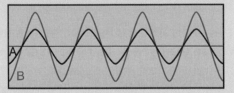

The trace shows two waves with the same frequency. Wave A has half the amplitude of wave B.

Ultrasound travels in **ultrasonic waves**, beyond our range of hearing.
Echo sounding is used to measure distances by recording how long
a wave takes to travel back to its source after being reflected.
Bats echo sound with ultrasonic waves to avoid hitting objects.
Ultrasound echo sounding is used by geologists studying rock structure
and by ships' sonar, checking sea depth.
In medicine, ultrasound is used to image parts of the body, including
unborn babies in pre-natal scans. Ultrasonic waves are used to destroy
kidney stones and help broken bones to heal quicker.

The **seismic waves** produced by earthquakes are used to study the Earth.
Seismic **P waves** are longitudinal waves (com**p**ression waves).
Seismic **S waves** are transverse waves, often called **s**hear waves.
Both types of wave travel in solid rock, only P waves travel in liquid rock too.

Analogue signals have a wave amplitude that
keeps changing. They can also have a range
of different frequencies in one signal.
Recorded speech is an analogue signal.
Digital signals are pulses with only two possible
values, on or off.

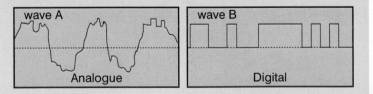

Telephone systems and many radio and TV broadcasts use digital signals.
They give a higher quality transfer and more signals can be sent down each cable.
Optical fibres are replacing wires in communication systems as they can transfer
far more information. They carry light or infrared rays as digital signals.

Questions

1 Copy and complete:

   i) Sound waves are ...... waves.
      They are caused by ...... and they need a medium to ...... through.

   ii) The frequency of a sound wave is the number of vibrations per ...... 1000 Hz is 1 ......, a ...... Hz is 1 MHz.

   iii) The ...... of a sound wave is its loudness.
       Loud sounds transfer ...... energy than ...... sounds.
       Loudness is measured in decibels (......).

   iv) Ultrasound is ...... frequency sound waves beyond the ...... range of hearing.
      Ultrasonic waves are used in ...... sounding by bats, ships ...... and in rock analysis.
      Ultrasound scans are used for pre-...... baby checks. Ultrasonic waves can be used to break up ...... stones.

   v) Analogue wave signals have a range of am...... They can also have a selection of different fr......
      Digital wave signals have only two amplitude values, ... or off.

   vi) Digital signals are used in communication systems because they are ...... quality.
       Optical ...... are replacing wire cables because they can carry more ...... in the same thickness cable.

2 Look at these sound wave traces.

   They are from a cathode ray oscilloscope.

   i) Which wave has the higher frequency?

   ii) Which wave has the larger amplitude? Both waves are travelling at the same speed.

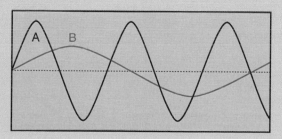

3 a) Put the following sounds into order of increasing amplitude (loudness).
     A road drill (100 dB),
     car engine ticking over (50 dB),
     noisy class of year 10 students (80 dB),
     water moving in a heating pipe (15 dB).

   b) Decide where you should add these sounds into the loudness order you worked out above.
      A Jumbo jet taking off, watched nearby,
      a cat creeping up on a mouse,
      a teacher shouting at a Year 10 class.

4 Fishermen use sonar to find fish.
  Draw a sketch of a fishing boat using echo sounding to locate a shoal of fish.
  Draw the ultrasonic waves coming away from the boat with arrows. Show some of the waves being reflected off the fish and returning to the boat.

5 Ultrasonic waves travel at the same speed as audible sound waves in air.

   i) Do they have shorter or longer wavelengths than audible sounds?

   ii) Do they have higher of lower pitch than audible sounds?

6 Most televisions can only receive analogue signals, but:
   - Mini satellite dishes only receive digital signals.
   - DVD players read digital information off the DVD discs.
   - Both of these can be connected up to a normal TV.
  Find out how digital transmissions can be watched on a normal analogue TV.

CHAPTER 15

The electromagnetic spectrum

▶▶▶ 15a Looking at electromagnetic rays

Electromagnetic rays affect you every second of every day.
When you are awake, you see because of light.
You radiate infrared rays from your body.
You are continually being hit by radio waves, microwaves
ultraviolet rays and some traces of gamma rays.
You can't escape from electromagnetic rays.

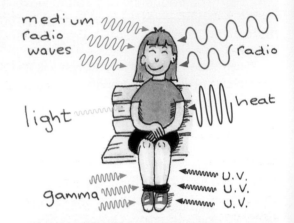

The order of the electromagnetic spectrum

Step 1: *starting with visible light*

a) i) What is the following saying used for?
    Richard of York gave battle in vain:
  ii) What does **ROY G BIV** stand for?

You have probably known the order of the colours
of the rainbow since you learnt them at primary school.
Pure white light contains all the colours.

b) Look at the picture of light travelling through a prism.
    Which colour of light is refracted (bent) the least?

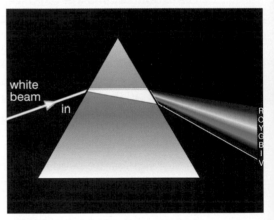

When a beam of white light enters a glass prism from air, it slows down.
All the light is refracted (see page 173). Red light is refracted
less than violet light because it travels faster in the glass.

Step 2: *putting infrared and ultraviolet into order: lowest to highest energy*

Next to which visible light colours would you expect to find:

c) Infrared rays?   **d)** Ultraviolet rays?

The order so far

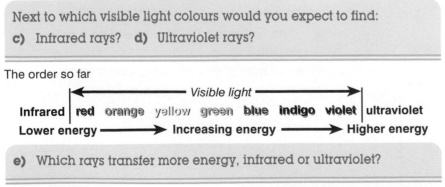

e) Which rays transfer more energy, infrared or ultraviolet?

Infrared rays are safer than ultraviolet rays. We use infrared rays
in a conventional oven for cooking. As long as we avoid high temperatures
infrared rays are safe. Ultraviolet rays transfer more energy
than infrared rays. They can cause skin cancer and damage our eyes.

Step 3: *adding microwaves and X-rays to the spectrum*

X-rays are more dangerous than microwaves. We use microwaves in cooking and in mobile phones. People who use X-rays in hospitals always use lead barriers to protect themselves from being hit by them.

Mobile phones use microwaves.

> **f)** X-rays go on one side of the spectrum, microwaves on the other. To which side would you add each wave?

X-rays are more dangerous than ultraviolet because they transfer more energy. Microwaves transfer less energy than Infrared rays. Adding in X-rays and microwaves we get:

microwaves	Infrared	visible light	ultraviolet	**X-rays**
────────── Increasing energy being transferred ⟶				

Step 4: *placing radio waves and gamma rays*

Gamma rays are **very high energy** rays released from the nucleus of radioactive elements (see page 204). High exposure to gamma rays can cause damage to body cells and can lead to death. Radio waves are used for communications everywhere in the world, they have long wavelengths and are safe to use. When we add them into the spectrum, we get:

long   wavelength   short

radio waves / microwaves / infrared / visible light / ultraviolet / X-rays / gamma rays

low   frequency   high

low   energy   high

As the wavelength of the rays in the electromagnetic spectrum decreases the frequency increases.

> All electromagnetic rays (including light) travel at 300 million metres per second in a vacuum.

Remind yourself!

1 Copy and complete:

The waves that make up the ...... magnetic spectrum are: ...... waves, microwave, ... red rays, visible light, ... violet rays, X-......, and gamma waves.
All these ...... travel at the speed of ......
Pure ...... light contains red, orange, ..., green, ..., indigo and ...... light.
When a ray of white light splits in a ...... the purple light is refracted more than the ...... light.

2 Invent a way of remembering the order of the rays of the electromagnetic spectrum.

Try another rhyme or maybe a rap.

3 When an electromagnetic wave changes speed, its frequency does not change. What happens to the wavelength of a ray of light when it slows down?

(Hint: look at equation 14, $v = f \times \lambda$)

Have you ever tried tuning in a radio?
As you adjust the frequency, you pick up many different radio stations.
Turning off the radio might stop the sound but what happens
to all those radio waves? Where do they go? Are they still there?

While you are reading this page, your body is absorbing thousands
of radio and TV broadcasts. Hundreds of mobile phone text messages,
weather forecasts and gripping soap operas are passing straight
through your head. It's a good thing we can't hear it all!

The sound and light in a TV broadcast are both transmitted together
as a radio wave or microwave. The TV then splits the two up to give
a picture on the monitor and sound from the speaker. You would only
get a delay in the sound if you went half way down the road to watch.
(You'd need the sound really loud and a pair of binoculars!)

a) Light travels a million times faster than sound.
Explain why voices always match people's lips moving in TV programmes.

When electromagnetic waves hit a conductor, they induce a small alternating
current in it. The frequency of the a.c. current will be the same
as the frequency of the wave. This is how aerials and satellite dishes
are able to receive TV and radio signals. When you tune a TV or radio,
you are really selecting one of the many tiny currents induced
at the aerial to be amplified.

b) Explain what happens inside a radio when you tune it.

satellite dishes

Radio wave transmissions

Many local radio and TV programmes are transmitted to our homes
from local radio towers. They are normally placed on high ground
so that they can cover a large area. Their range is limited because of
the curvature of the Earth.

c) What effect does it have on the quality of TV reception if the
aerial is not pointing at the transmission tower?

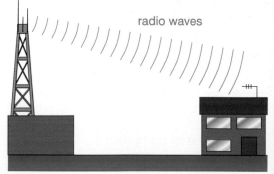

radio waves

Longer wavelength radio waves can travel further around the Earth.
They are more easily diffracted so they spread better
than shorter waves (see pages 174–175).
Long wave radio waves are also being **reflected** by the **ionosphere**.
It is a layer of charged particles high in the Earth's atmosphere.
Waves hitting it are reflected on further around the Earth's surface.

ionosphere

radio waves

Microwave transmissions

Some frequencies of microwaves can penetrate
the Earth's atmosphere. Narrow beams of microwaves
are used to send TV and radio signals up to geo-stationary
satellites above the equator (see page 155).
These satellites reflect the signals back to the Earth.
Just one satellite can reflect TV signals all over the UK.

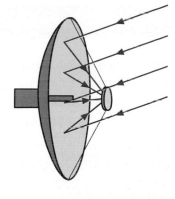

d) Why do all the satellite dishes in the UK point south?

Radar systems used to track planes and ships use reflected microwaves.
They work a bit like a simple sound echoing off a wall. The radar beam
from the transmitter hits the plane and bounces off. Any beam
reflected back towards the radar station is recorded and displayed on the radar screen.

Microwaves are used by the mobile phone networks. Over recent years,
hundreds of new radio masts have been erected.
Your mobile phone is in continual contact with your network via
the closest mast. If you are travelling, the signal strength varies
as you move between masts. The control system monitors which mast
will reach your phone if you have a call.

Fitting a satellite dish.

e) Are there any health risks from using a mobile phone too much?

Microwaves and water

Research is being carried out to find out exactly what effect any heat
released by microwaves from mobile phones has on the brain.

Any electromagnetic ray releases some heat when it is absorbed.

Certain frequencies of microwave are easily absorbed
by water molecules. These microwaves cause the water
molecules to vibrate violently. Cells contain large amounts
of water so the heat released can damage or kill them.
This is exactly how a microwave oven cooks food.
They heat only the water contained in the food directly and quickly.

Remind yourself!

1 Copy and complete:

We transmit radio and ...... signals as either ......
or ...... waves.
Long wave radio ...... are reflected off the
ionosphere, a layer in the ...... atmosphere.
Some ...... of microwave are reflected off geo-
...... satellites. Microwaves are also used by
mobile ...... and in microwave ......

2 Pupil survey: design a simple survey sheet to
help you find people's feelings about mobile
phones. Here are some hints:

i) Have they noticed one get hot?

ii) Do they give people headaches?

iii) Should they be used 'hands free'?

iv) What do people think about the microwaves
in the air because of mobile phones?

Have you ever had an argument about which station to watch on TV?
You might want a music show while someone else
in your family wants the football. First you fight over who holds
the remote control. If you lose that fight, you go to the TV
and change the channel. Then you sit there with your hand
over the **infrared** sensor to stop the remote from working.

a) Infrared rays can penetrate skin. Why does putting your hand
in front of a TV sensor panel stop the remote control from working?

Infrared rays

Infrared rays, (infrared radiation) are heat-carrying electromagnetic waves.
When you stand near a fire, the heat you feel is the infrared rays
penetrating your skin.

We first looked at the properties of infrared rays on pages 40–41.
- The hotter an object is, the more infrared rays it radiates (gives out).
- When infrared rays are radiated or **emitted** (given out) by a material,
  its atoms and molecules lose internal energy, so they vibrate less.
- When infrared rays are **absorbed** (taken in) by a material,
  its atoms and molecules gain internal energy so they vibrate more.
- Good radiators of infrared rays are also good absorbers (dull surfaces).
- Good reflectors of infrared rays are poor absorbers (shiny surfaces).
- Infrared rays travel at the speed of light in a vacuum.

b) Which classroom board would absorb more infrared radiation;
a black board or a shiny white board?

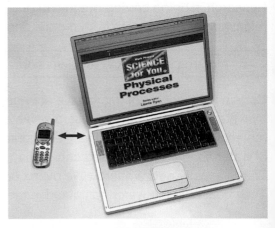

Uses of infrared rays

As well as being used by TV remote controls, infrared rays are also
used to control video recorders, DVD players and stereo systems.
Most mobile phones now have an infrared port that can receive
and send infrared rays. They are used with laptop computers
to provide totally portable access to the Internet. Mini disc players
can *download* music from a computer in a similar way.

Describe how infrared rays can be used to:
c) Grill some sausages, **d)** Toast some bread.
e) Heat a room with a heater using a shiny reflector.
(Hint: refer to the properties list above for extra help.)

Visible light and total internal reflection

Have you ever noticed on hot sunny days how objects near
to the ground can look blurred?
Sometimes it is possible to see mirages where the road
seems to disappear. It can look like a lake instead.
Sunlight gets refracted by the hot air just above
the road surface. Instead of reflecting off the road,
the light reflects off the hot air before travelling on to your eye.
You see a reflection of the sky instead of the road surface.
This is an example of **total internal reflection** of light.

Mirage: The plane's image seems to reflect off the runway.

f) Where have you seen the mirage effect?

Total internal reflection can also happen when a ray of light tries
to cross from glass back into air. If the angle at which the light
hits the boundary is too large, (above the **critical angle**)
the light will be reflected back into the glass again.
Total internal reflection can happen at any boundary
between two transparent mediums.

Cats' eyes in the middle of the road or any other type of reflectors use
total internal reflection to send rays of light back the way they came.

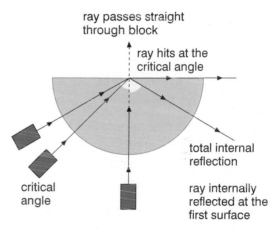

ray passes straight through block

ray hits at the critical angle

total internal reflection

critical angle

ray internally reflected at the first surface

Optical fibres are used to transmit light or infrared rays
by repeated total internal reflection. Telephone or video signals
converted into light or infrared, are sent in at one end of the fibre.
They travel to the other end of the fibre by bouncing off its inner surfaces.
An optical fibre can carry far more information than the same sized
electrical cable. The signals remain clearer as well.
Because optical fibres are flexible, they are used in endoscopes.
This device allows doctors to carry out internal examinations.
They have helped to revolutionise surgical techniques by providing
images inside the body during **keyhole surgery**.

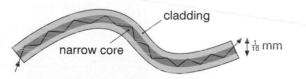

cladding

narrow core

$\frac{1}{10}$ mm

Remind yourself!

1 Copy and complete:

Infra...... rays are heat waves. They are used in
...... controls for TVs and ......
Grills and ...... use infrared rays.
Visible light and infrared rays can be used to
send information down ...... fibres.
They travel by repeated total ...... reflection. The
beam is ...... off the sides of the optical fibre as it
passes along it.

2 Design a reflector using a prism.

Show the path of the light as it passes through
the reflector and is sent back the way it comes.

3 Find out more about how endoscopes work and
what other uses they have.

ICT

Do you feel happier on a sunny day?
Many people find that their mood is much better when
there is plenty of sunlight about. Some people even sit in front
of a special light every day during the winter to get more light.
Maybe that's why we take holidays 'in the Sun'.

It is not clear why sunlight affects some people in such an extreme
way, but we need some exposure to the Sun.
The ultraviolet rays present in sunlight help our skin to make
vitamin D. This helps to keep our bones and joints healthy.

Ultraviolet rays give us a tan but can damage your eyes and cause skin cancer. The damage to skin cells on your face causes premature ageing.

Ultraviolet rays

We might need ultraviolet rays to help us to have enough vitamin D
but too much is very bad for our eyes and skin. Over-exposure to
ultraviolet rays can kill cells or cause skin cancer.

The problem is that people like to have a sun tan, it makes them feel
more healthy and attractive. Whenever you go into strong sunlight it is vital
that you protect any exposed skin with a suitable sun protection cream.
Good quality sun creams should protect your skin from the most harmful
frequencies of ultraviolet light, while still helping your skin to tan a little.

a) What factor sun cream do you normally use?

People with darker skins absorb less ultraviolet than those
with lighter skin. If you have a fair complexion, you should use a high
factor sun cream and stay in the Sun for less time.
Luckily for us, the ozone in the Earth's atmosphere cuts out most
of the more dangerous frequency ultraviolet rays.
In Australia and New Zealand people have to be extra careful
in protecting their skin because there is less ozone in the atmosphere.

Sun beds are used to help people to get a quick tan.
They use special fluorescent lamps that give off ultraviolet light.

b) Why do people have to wear sunglasses when they use sun beds?

We use ultraviolet light in discos as part of the light show.
It makes white clothes fluoresce (give off bright visible light).
Security markings are often made with invisible inks that only show
up when we shine ultraviolet rays on them.

Alterations and forgeries in paintings can be detected using ultraviolet light.

X-rays

Have you ever seen an X-ray of your teeth?
You see your gums as a faint trace while your teeth
are clearly outlined. Any metal based fillings show up like solid
pegs while cavities are dark holes in places they shouldn't be!

It's quite possible that you will have had an X-ray to check for
broken or fractured bones, or maybe as part of a medical examination.

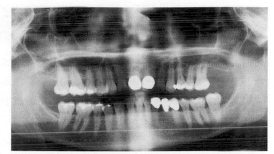

> X-rays are high energy electromagnetic waves.

They easily pass through soft body tissue and only get absorbed
by bone or any metals. This is why an X-ray picture will only show
faint traces of any body organs while bones are clearly visible.

> **c)** How many times have you had an X-ray?
> Compare your answer with your partner's (don't forget the dentist).

If you have had an X-ray, you probably noticed that the person who took
it moved behind a protective screen. Everybody else normally
leaves the room to avoid unnecessary exposure to the rays.
But you can't feel anything happening anyway!
You might not be able to feel anything but you must respect X-rays.
People are very careful with X-rays because over-exposure
to them can kill or damage cells, possibly leading to cancer.

Gamma rays

The electromagnetic rays with the most energy are gamma rays.
These are one of the three different types of **radiation** that are
released from the **nucleus** of **radioactive** atoms.
Gamma rays can also kill or damage cells, leading to cancer.
They will kill bacteria so gamma rays are used to sterilise surgical equipment.
The properties and further uses of gamma rays are described in Chapter 16.

> **d)** If you were going to have an operation, would you prefer they prepared
> the surgical tools with gamma radiation or just boiled them first? Why?

Remind yourself!

1 Copy and complete:

Ultra...... rays are present in sun......
They cause skin to ...... as they are absorbed.
Ultraviolet rays help the skin to produce ...... D.
Over-exposure to them can ...... cells or lead to
skin ......
X-rays are used to examine the inside of ......
They penetrate soft tissue but are absorbed by
...... or metal. X-rays and the high energy
gamma ...... can also kill ...... or cause cancer.

2 Which of the following makes use of ultraviolet
and which of X-rays?

i) Baggage security check at an airport;

ii) washing powders that make white clothes
very white;

iii) machines that check bank notes in shops;

iv) dentists checking for tooth decay;

v) attracting insects into traps;

vi) sun beds to give quick tans.

193

Summary

The rays in the **electromagnetic spectrum** all travel
at the speed of light in **free space**, normally called a **vacuum**.
They are listed below, in order of longest to shortest wavelength:

long ⟵ wavelength ⟶ short

radio waves / microwaves / infrared / visible light / ultraviolet / X-rays / gamma rays

low ⟶ frequency ⟶ high

low ⟶ energy ⟶ high

Radio waves and **microwaves** are used for transferring TV and radio signals.
Long wave radio waves travel around the Earth by being reflected
by the **ionosphere** (a layer of charge particles in the atmosphere).
Some frequencies of microwaves can penetrate the Earth's atmosphere.
They are used to reflect TV and radio signals off satellites back down to Earth.
Microwaves are used by mobile phones and also to cook food in microwave ovens.

Infrared rays transfer heat energy when they are absorbed.
Any object cooling will radiate out infrared rays.
We use infrared rays to work TV, video and DVD player remote controls.
Many mobile phones and computers can communicate using infrared rays.
Toasters and ovens cook using infrared rays and fires give them out.

Total internal reflection happens at boundaries between different mediums.
A ray of visible light or infrared gets reflected back instead of crossing
the boundary if the ray meets the new medium at too large an angle.
Optical fibres transfer light or infrared rays by repeated total internal reflection.
Telephone and video signals are transferred by optical fibres.
We also use them in endoscopes during **keyhole surgery**.

critical angle

total internal reflection

We need **ultraviolet rays** to help our skin make vitamin D.
Over exposure to ultraviolet rays can kill cells or cause skin cancer.
People with darker skin absorb less ultraviolet than those with lighter skin.
We use ultraviolet fluorescent lamps in sun beds.
Security markings use invisible inks, which only show up in ultraviolet light.

X-rays and **gamma rays** are the electromagnetic rays that transfer
the most energy. Over-exposure to either can kill or damage cells.
Damaged cells can start to form cancerous growths.
X-rays are used for internal examinations because they penetrate
soft tissue very easily but are absorbed by bones.
Gamma rays kill bacteria so they are used to sterilise surgical equipment.

Questions

1 Copy and complete:

i) All the rays of the electromagnetic spectrum travel at the speed of ...... in free space which is called a v......

ii) Listed in order from longest to shortest wavelength the ...... rays are: Radio waves, ..., infrared rays, visible light, ...violet rays, X-rays and ...... rays.
Gamma waves have the highest f...... so they transfer the most ......

iii) Long wave radio waves are reflected around the world by the ......
Narrow beam ...... are reflected off geo-stationary ......

iv) Infrared rays are used in TV and video ...... controls. They are also used in ...... to cook food.

v) ...... fibres pass infrared and visible light by repeated total ...... reflection.

vi) Ultraviolet ...... are present in sunlight. They help our skin produce vitamin ...... Over-exposure to ultra...... rays can kill cells or cause skin cancer.

vii) X-rays can penetrate soft tissue but are absorbed by bones. Gamma rays are used to kill ...... on surgical instruments.
Both X-rays and gamma rays can kill cells and cause ...... growths.

2 When white light is split up by a prism:

i) Which colour light is refracted the least?

ii) Which colour light is refracted the most?

iii) Red light travels fastest through the glass prism, which light travels the slowest?

iv) The wavelength of violet light is the shortest inside a prism.
Which colour of visible light has the shortest wavelength in air?

3 Create a chart or poster to show the rays of the electromagnetic spectrum in order.
Add arrows to show the direction of increasing frequency and wavelength.
Highlight the high and low energy ends.
Then, for each type of electromagnetic ray, add sketches showing some common uses and any dangers.

4 Night vision cameras are used in helicopters by the police to track stolen cars and follow escaping criminals. These cameras detect heat and convert it into visible light images.

i) What type of electromagnetic rays do they receive and turn into visible light?

ii) Which part of the body do you think radiates the most heat?

iii) Explain why it is very difficult to hide from these cameras even in complete darkness.

5 Sketch a light ray travelling down an optical fibre. Explain in your own words why the ray of light does not just pass through the sides of the fibre.

6 Imagine you were setting off on holiday to a hot Mediterranean country with a group of friends. Some of them said it was very stupid to bother with sun protection cream. How would you persuade them to use a cream and avoid the risk of skin cancer.

7 When you have an X-ray taken of your stomach and kidneys, they often place a very heavy lead cover over your rib cage. Why do they cover that part of your body?

Radiation can be very dangerous.
High exposure will kill and there is no cure.
So why do we:
- *use it in smoke detectors?*
- *inject radioactive sources into people in hospitals?*
- *show radioactive samples to classes of students?*

These are examples of radiation being used carefully.

Background radiation is everywhere so exposure is impossible to avoid.
Even so, most people never become exposed to any dangerous levels.
The amount of radiation used during medical treatments is carefully
controlled by specialist staff. They make sure that risks are small.
People who work with radiation do have to watch the amount
they are exposed to but they are always very cautious.

a) What is meant by background radiation?

Radiation and ionisation

Radiation causes neutral atoms or molecules to lose or gain electrons.
They become charged particles called ions (see pages 80–81).
We say that the atoms or molecules have been ionised.
An alpha radiation source will emit a stream of alpha particles.
They do not travel very far in air but can produce between
1000 and 10 000 new ions every millimetre as they travel through air.
A stream of beta particles from a source might only create
100 ions per mm but beta particles do travel further than alpha.
Gamma rays create far fewer ions than the other types of radiation.
But they are very difficult to stop so they can still be very destructive.

b) Which type of radiation creates the most ions as it passes through air?

The effects of radiation on living cells

Radiation passing through a living cell can kill or seriously damage it.

Atoms and molecules belonging to the chemicals inside our cells
can be ionised. If parts of the DNA molecules within a cell become
ionised, the genes in the chromosomes get altered.
Changing DNA can damage sex cells (the egg and the sperm)
or result in deformed children. Changing a cell's structure
can make a cell more likely to produce cancer cells when it divides.
Cancer cells multiply in an uncontrolled way forming a tumour.

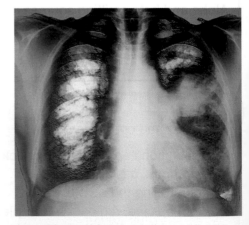

The righthand lung showing damage from cancer tumours.

c) If the egg cells in a woman's ovaries are damaged, can she still have a baby?

Absorbing radiation into your body

The risks of radiation sources near to your body vary depending upon which type of radiation the source releases.
If you breathe in radioactive dust or eat radioactive food, there is a much higher chance of internal damage to your body.

Alpha radiation can only pass through a few centimetres of air. It is stopped by thin paper and skin. This means that outside the body, alpha radiation is reasonably low risk. If it can't pass through your skin, it is unlikely to reach many living cells.

Alpha particles released from a source inside your body are far more dangerous! They cause the most ionisation so are likely to kill or damage the cells of living tissue.

Beta and gamma radiation are more dangerous than alpha outside the body because they can penetrate skin. Once inside the body they will cause ionisation affecting the cells of living tissue. Beta and gamma ionise less than alpha but are more able to reach anywhere inside the body as they are harder to stop.

Alpha particles can't get through skin. If they do get inside the body they cause damage to cells.

d) Why aren't alpha sources used inside patients?

Working with radiation

Radiation is used in many industries as well as in medicine and nuclear power stations. Any people who work with radiation have to avoid too much contact at any one time. They often wear badges that monitor how much radiation they have been exposed to. The badges use photographic film which gets darker with exposure to radiation. If the developed film from a worker's badge is too dark, they will be assigned to a job with a lower contamination risk.

radiation badge

▶▶▶ 16d The atom

Everything is made out of atoms.
They are so small that we can't see them!

Atoms contain even smaller subatomic particles.
There are protons and neutrons inside the nucleus of atoms.
Electrons orbit around the nucleus.
It gets better. Not only can we not see atoms, we know that there is something smaller inside them!

Neutrons have no charge, protons have a positive charge and electrons have a negative charge.
So we can measure their charge, even though we can't see them!

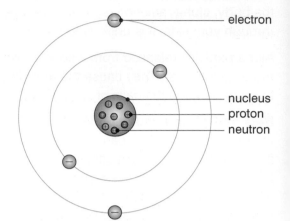

a) Is the structure of the atom outlined above and shown in the sketch above a fact or a theory?

Discovering the atom

Scientists have not always had a clear idea about atomic structure.
At the end of the 19th century, atoms were thought to be tiny particles of positively charged matter with negative electrons mixed in.
This was the plum pudding model of atoms.

The present day theory comes from an experiment carried out in the early years of the 20th century by Ernest Rutherford working with a team of physicists. They wanted to prove that the plum pudding model for the structure of the atom was correct; they didn't!

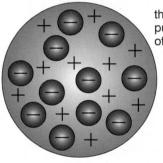

the plum pudding model of the atom

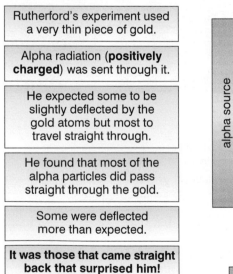

Rutherford's experiment used a very thin piece of gold.

Alpha radiation (**positively charged**) was sent through it.

He expected some to be slightly deflected by the gold atoms but most to travel straight through.

He found that most of the alpha particles did pass straight through the gold.

Some were deflected more than expected.

It was those that came straight back that surprised him!

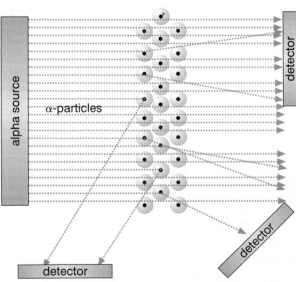

Rutherford decided that the positive alpha particles must be being repelled by a strong positive charge.

He concluded that the gold atoms were not solid lumps as the plum pudding model suggested.

Our present theory is that the **nucleus** is at the centre of an atom.

It contains most of the mass and all the **positive** charge.

The rest of the atom is mostly empty space with electrons orbiting the nucleus.

b) What made some of the alpha particles bounce back the way they came? (Hint: like charges repel.)

c) Why did most of the alpha radiation just pass straight through?

Atomic number and mass number

The table on the right shows the relative mass
and charge of the subatomic particles.
They are used to compare atoms with other atoms.
The true masses of neutrons and protons are very small
(approximately 0.000 000 000 000 000 000 000 000 0017 kg).

Subatomic particle	Relative mass	Relative charge
Neutron	1	0
Proton	1	+1
Electron	Negligible	−1

The **atomic number** (or **proton number**) tells you how many
protons an atom has. All the atoms of a certain element
will have the same number of protons.

> Changing the number of protons in an atom gives a completely different element.

Neutrally charged atoms have the same number of electrons and protons.
For example; neutral oxygen atoms have 8 protons and 8 electrons.

> **d)** The atomic number of neon is 10. How many protons does it have?
>
> **e)** An oxygen ion has a charge –2. How many extra electrons will it have?

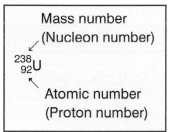

Mass number
(Nucleon number)

$^{238}_{92}U$

Atomic number
(Proton number)

Protons and neutrons are often called nucleons because
they are found inside the nucleus.

> The mass number (or nucleon number)
> is the number of neutrons plus the number of protons.

Isotopes

Isotopes are atoms that have the same atomic number
but different mass numbers. They are from the same
element but they have different numbers of neutrons.

> Chlorine has two very common isotopes; ^{35}Cl and ^{37}Cl.
> The isotopes have 17 protons; (atomic number of Cl is 17).
> The isotope ^{35}Cl has 18 neutrons (35 − 17 − 18)
> ^{37}Cl has 20 neutrons (37 − 17 = 20).

The table on the right shows some more elements
with their most common isotopes.

Element	Atomic number	Mass (Nucleon) number	% in a typical sample
Lithium	3	6	7.5
Lithium	3	7	92.5
Carbon	6	12	99
Carbon	6	13	1
Uranium	92	235	99.3
Uranium	92	238	0.7

Remind yourself!

1 Copy and complete:

Protons and ...... are found inside the nucleus of
atoms. Electrons orbit the ......
The ...... number of an element is the number of
protons. The ...... number is the number of ......
and the number of ......
Elements can have different i...... which are
atoms with the same ...... number but a different
...... number.

2 Use the isotope table above to answer these
questions:

i) Which isotope of lithium is most common?

ii) How many protons do the carbon isotopes
have?

iii) How many neutrons do each of the uranium
isotopes have?

iv) What mass of carbon-13 is there in 1 kg of a
typical piece of pure carbon?

▶▶▶ 16e Half-life and radioactive decay

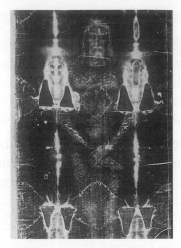

Look at the picture opposite of the Turin Shroud:

The true age of the Turin Shroud was found by **carbon dating**. Cosmic rays hitting carbon in the atmosphere make small numbers of carbon-12 atoms change into another isotope called carbon-14. Plants and trees use carbon in the process of photosynthesis. We make cloth from cotton plants so in any newly made cloth, tiny amounts of carbon-14 are present.

Carbon-14 is an example of a **radioisotope**, a radioactive element. Its nuclei are unstable, they **radioactively decay**.

Carbon-14 radioactively decays turning back into carbon-12.

In 5570 years, half of the Carbon-14 atoms in a cloth sample will have decayed. Comparing the **decay rate** from an old cloth to the rate from a similar new cloth gives an age for the older cloth.

For hundreds of years it was believed that the Turin Shroud was the cloth used to cover Jesus after he was taken down from the Cross. It is stored in a vault in Turin Cathedral in Italy. We now know that the cloth it is made from is only 600 years old, while Jesus was crucified 2000 years ago.

a) Imagine a cloth made 5570 years ago contained 1000 carbon-14 atoms. How many would it have now?

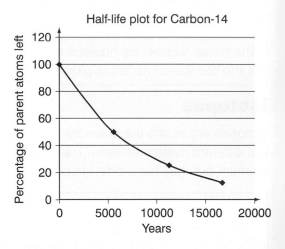

We use radioactive dating to find the age of rocks containing uranium.

A decay equation for Uranium:

$$^{238}_{92}U \longrightarrow \text{gives off alpha } ^{4}_{2}He^{2+} \longrightarrow ^{234}_{90}Th \text{ (thorium is made)}$$

When an atom of a radioisotope decays, a different element is formed: This table shows the effect on an atom when it emits radiation.

Type of radiation	How parent radioisotope changes		Different element formed?
Alpha	Atomic number drops by 2	Mass number drops by 4	Yes
Beta	Atomic number goes up by 1	Mass number no change	Yes
Gamma	Atomic number no change	Mass number no change	No

All radioactive materials become less radioactive as time passes. The number of **parent** atoms decaying by giving out radiation decreases.

The time it takes for half the parent atoms to decay is called the **half-life**.

When a radioactive material loses its radioactivity quickly we say it has a short **half-life**. Measuring the half-life of a sample involves counting the number of radioactive decays per second.

b) Find the half-life period for the sample shown in the graph.

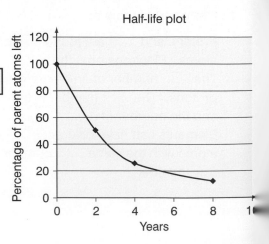

Summary

Alpha, beta and gamma are the three types of radiation.
They are released from the nucleus of **radioactive atoms** when they **decay**.
We are always being exposed to **background radiation**, which is given off
by certain rocks and soil. The food we eat, some medical treatments
and industrial processes also release small amounts of radiation.

Radiation causes **ionisation**. Neutrally charged atoms or molecules become
charged. Radiation absorbed into tissue can cause cells to be damaged or die.
Damaged cells can divide and multiply out of control, forming cancer tumours.

Property	Alpha (α)	Beta (β)	Gamma (γ)
Description of radiation	Helium nucleus with no electrons	Fast moving electron	High energy electromagnetic ray
Charge	+2	–1	No charge
Stopped by	Skin, paper or 5 cm of air	A few metres of air, 3 mm of aluminium	Thick concrete or lead.
Causes ionisation or not	Very ionising. 1000 to 10 000 new ions every mm	Some ionisation. Can create 100 new ions every mm	Only slightly ionising but can penetrate a long way
Speed	Travel at about 10% the speed of light	Travel at about 50% of the speed of light	Travel at the speed of light
Danger	Most dangerous inside the body	Dangerous from outside the body as they can penetrate the skin	Dangerous from outside the body as they can penetrate the skin

Radiation is used to check the strength of welds, monitor the thickness
of paper and metal sheets and to help smoke detectors work.
Some medical procedures use radiation to help to see what's happening
inside your body.
Radiation is also used to kill off cancerous tumours.
The time it takes for half the **parent** atoms from a radioactive source to
decay is called the **half-life**. Medical processes use gamma sources with
a short half-life. This ensures that patients are not over-exposed to radiation.

Atomic number (proton number) = number of protons.
Neutral atoms have the same number of electrons and protons.
Mass number (nucleon number) = number of neutrons + number of protons.
Isotopes are atoms that have the same atomic number
but different mass numbers (same element, different mass).
A **radioisotope** is a radioactive element with unstable nuclei.
They **radioactively decay** by giving off radiation.
If they give off alpha or beta radiation, the radioisotope nucleus
disintegrates. An atom of a different element is formed.

Subatomic particle	Relative mass	Relative charge
Neutron	1	0
Proton	1	+1
Electron	Negligible	–1

Questions

1 Copy and complete:

i) The three types of radiation are ......, beta and ......
They are released from the ...... of a radioactive atom when it ......
If radiation gets absorbed into the body, it can ...... or badly damage cells.
It might cause ......

ii) ...... radiation is everywhere.
It is released from rocks and from the ......, background radiation can also be released from food, medical ...... and many industrial processes.

iii) ...... radiation is a ...... nucleus.
It is a highly charged particle with no ......
Alpha radiation is stopped easily in air, by skin and ...... Inside the body it is the most ...... type of radiation.

iv) Beta radiation is high speed ......
It is less ionising than ...... but much more penetrating. It takes ...... mm of ...... to stop beta radiation.

v) Gamma radiation is not a ......, it is a high energy ...... magnetic wave.
...... radiation ionises less than ...... or beta but is much more difficult to ......

vi) Radiation is used in ...... to help with internal examinations and also to kill cancer.
Gamma sources with a short ......-life are used in medical processes, so that no ...... is left inside the patient for too long.

2 Use the information in the summary and on the previous page to define the following:

i) Atomic number (proton number).

ii) A neutrally charged atom.

iii) Ionisation.

iv) Mass number.

v) Isotope.

vi) Radioactive decay.

3 Read through Rutherford's experiment (see page 202).

i) What inside the nucleus of the gold atoms made some of the positive alpha particles rebound, travelling back to the source?

ii) How did the alpha radiation returning towards the source prove that the nucleus of an atom was not negatively charged?

iii) Why did so many alpha particles travel straight through the gold?
(Remember, they are stopped by paper.)

4 Carbon dating uses the fact that carbon-14 isotopes decay back to $_{12}C$ atoms.
The half-life of $_{14}C$ is 5570 years.

i) Imagine a piece of preserved wood had 1000 $_{14}C$ atoms. How many $_{14}C$ would it have had 5570 years ago?

ii) How many $_{14}C$ atoms should the wood have in 5570 years, if it is preserved?

iii) How long would it be before there are no $_{14}C$ atoms left?

5 Background radiation is everywhere;

i) Do you think people should worry about the radiation they get from eating their dinner?

ii) Would you think twice about digging in the garden now you know that you will be releasing radiation? If not, why not?

6 What length of half-life do you think the gamma sources used in medical treatments would have, short or long? Explain why.

7 When the nuclear power station at Chernobyl exploded, all the workers were exposed to life threatening doses of radiation. Military personnel were sent in to help prevent the radioactive materials from spreading.
What do you think the effect of the death of workers and the military people was on their families? What went wrong? Why?
Write a newspaper article about Chernobyl.

Further questions on Waves

1 The diagram shows some waves travelling along a rope.

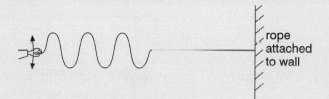

rope
attached
to wall

(a) Show on a copy of the diagram

    (i) the wavelength of one of the waves   (2)

    (ii) the amplitude of one of the waves   (2)

(b) The waves shown on the diagram were produced in two seconds.
What is the frequency of the waves?   (2)
(AQA (NEAB) 1999)

2 The diagram shows a ray of light travelling through a glass block.

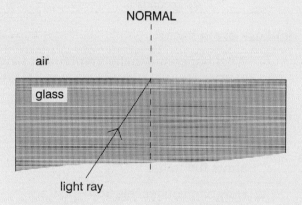

NORMAL

air

glass

light ray

(a) Copy and complete the diagram to show what happens to the ray of light when it comes out of the glass.   (2)

(b) Explain why this happens to the ray of light.
   (2)
(AQA (NEAB) 1999)

3 (a) Nicola connects a microphone to an oscilloscope.

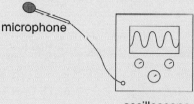

microphone

oscilloscope

She holds a vibrating tuning fork near the microphone.
The display on the oscilloscope screen is shown below.

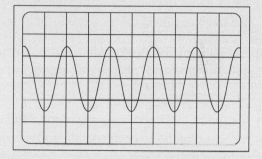

How does the display change:

  (i) when the tuning fork is held further away from the microphone;   (1)

  (ii) when Nicola uses a higher frequency tuning fork?   (1)

  (iii) Copy and complete the sentence below to describe the energy conversion taking place.
The microphone converts ...... energy into ...... energy.   (1)

(b) The diagram shows a pregnant woman being given an ultrasound scan.

  (i) What is ultrasound?   (1)

  (ii) Explain why ultrasound scanning is important in medicine.   (3)
(EDEXCEL 1999)

4 A pest control device emits sound at frequencies between 50 kHz and 70 kHz.
The device is shown in the diagram.

Pest Control™

ELECTRO-MAGNETIC®

Magnette-Sonic

HARMLESS TO CATS, DOGS AND PEOPLE

BUT RATS, MICE AND ANTS CAN'T STAND IT

GUARANTEED
to rid any building
of Rats, Mice, Ants
and Spiders

(a) What name is given to sound with a frequency greater than 20 kHz? (1)

(b) Explain why the device affects rats, mice and ants but does not affect cats, dogs and people. (3)

(c) A speech therapist uses a balloon filled with carbon dioxide to concentrate sound waves onto a person's ear.
Sound is transmitted in the form of waves.
Sound travels more slowly in carbon dioxide than in air.
The diagram shows what happens to the sound waves.
Explain what happens to the waves as they travel from the source to the person's ear.

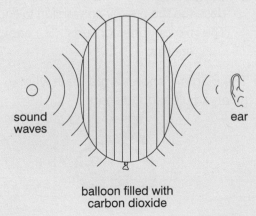

sound waves

ear

balloon filled with carbon dioxide

(3)
(EDEXCEL 2000)

5 This question is about waves.
The waves made by earthquakes are recorded by an instrument called a seismometer.

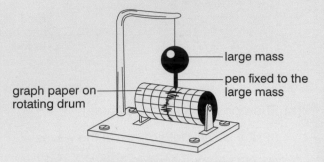

large mass

pen fixed to the large mass

graph paper on rotating drum

Here is a trace of the waves recorded by the seismometer after an earthquake.

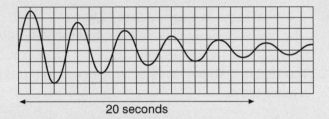

20 seconds

(a) Look at the trace.

(i) What happened to the **amplitude** of the earthquake wave during the 20 seconds? Select the correct answer from the words below.

**decreases   increases
stays the same** (1)

(ii) How can you tell this from the trace? (1)

(b) Use the information on the diagram to calculate the **frequency** of the recorded wave.
You **must** show how you work out your answer. (3)
(OCR 2001)

Further questions on Waves

6 (a) Copy and complete the sentences.
Sound waves are produced when objects ......
The greater the ...... of a sound wave the
louder the sound. (2)

(b) Electronic systems can be used to produce
ultrasonic waves.

  (i) What are *ultrasonic* waves? (2)

  (ii) Give **two** uses of ultrasonic waves. (2)
  (AQA 2001)

7 Steven has a compact disc (CD) player.

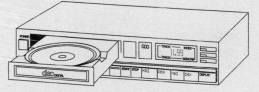

(a) Name ONE useful form of energy produced
from electricity by the CD player. (1)

(b) The CD player has a remote control system.
Name the type of electromagnetic wave used
in this system. (1)

(c) When Steven looks at the surface of his CD
discs he sees the colours of the spectrum:

red   orange   yellow   green
blue   indigo   violet

These colours are also produced by the
refraction of white light in a prism. The
diagram shows the refracted rays for the two
colours red and violet.

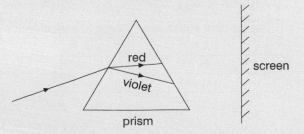

  (i) Draw, on a copy of the diagram, the light
  rays from the prism to the screen. (1)

  (ii) Give an example in which refraction is
  useful. (1)
  (EDEXCEL 1998)

8 The electromagnetic spectrum is the name given to
a family of waves that includes light, infrared and
ultraviolet radiations. All members of the family can
travel through a vacuum with the same high velocity.
Electromagnetic waves are produced when the
energy of electrically charged particles is changed
in some way. The greater the change in energy,
the shorter the wavelength of the electromagnetic
wave produced. Radio waves, with a wavelength of
up to 10 km and gamma (γ) rays with wavelengths
of a thousand millionths of a millimetre are found at
opposite ends of the electromagnetic spectrum.

(a) Name **one** part of the electromagnetic spectrum

  (i) that is **not** mentioned in the above
  passage, (1)

  (ii) that has a wavelength shorter than that of
  visible light. (1)

(b) Give **one** reason why radio waves have a
longer wavelength than gamma (γ) rays. (1)

(c) State **one** practical use of

  (i) infrared radiation, (1)

  (ii) ultraviolet radiation. (1)
  (WJEC 1999)

9 The boxes on the left show some types of
electromagnetic radiation.
The boxes on the right show some uses of
electromagnetic radiation.
Put this information into a table linking each
radiation to one of its uses.

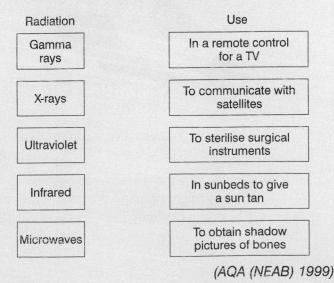

(AQA (NEAB) 1999)

10 (a) The diagram shows an atom (not drawn to scale).

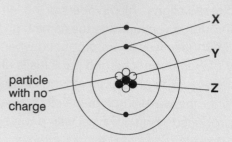

particle with no charge

(i) Use words from the list to label, **X**, **Y** and **Z** on a copy of the diagram.

electron   neutron
nucleus   proton   (3)

(ii) Select, from the list below, the mass (nucleon) number of this atom

3   4   7   10   (1)

(b) An isotope of iodine ^{131}I, has unstable nuclei. This radioactive isotope is used to investigate the circulation of the blood. The diagram shows a doctor measuring the amount of radioactive iodine moving through a patient's leg.

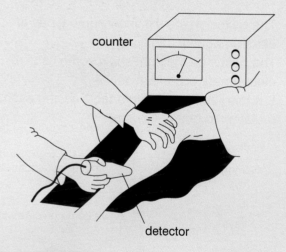

counter

detector

(i) Explain why radioactive iodine can be detected.   (2)

(ii) Explain why the use of a radioactive isotope might damage the patient unless it is strictly controlled.   (2)

(c) The diagram shows doctors examining an X-ray photograph of a patient.

Explain why X-rays can be used to take photographs of the inside of the body.   (2)

(AQA (NEAB) 2000)

11 The diagram shows what happens to the radiation from three radioactive substances when different materials are put in the way.

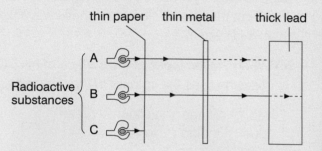

thin paper   thin metal   thick lead

Radioactive substances

A
B
C

Choose types of radiation from this list to complete a copy of the table below.

α (alpha)   β (beta)   γ (gamma)
UV (ultraviolet)

Radioactive substance	Type of radiation it emits
A	
B	
C	

(3)

(AQA (NEAB) 1999)

GCSE exams have 20% of the marks awarded for coursework.
Your teacher has to assess your practical skills.
You are given marks in 4 areas:

P   **Planning** to collect evidence.

O   **Obtaining** the evidence.

A   **Analysing** your evidence and drawing conclusions.

E   **Evaluating** your evidence.

Your teacher will mark you against checklists of points
to look out for in your work. You can see these for yourself
in the sections that follow.
Try to cover all the points, if they apply to your task, working
from 2 marks upwards.

Practical work is important!

P   Planning

Choosing apparatus

It is important to use the most suitable equipment.
For example, if you were calculating the speed of the fastest
runner in your class while they ran around the sports field,
you should use a stopwatch to measure the time. It would
not be a good idea to use a normal watch. Why not?

Deciding on how many readings

You will need to think about how many measurements
or observations to make in your experiments.
If you plan to show your results on a line graph,
aim to collect 5 different measurements.
And if the measurements are tricky to make,
you should repeat them. Taking the average of
the measurements will make them more reliable.

Safety

You should describe how you will avoid acidents in the test
you are planning. For example, if you were using heating equipment
as part of an insulation investigation, you would need to explain the
safety precautions you would use to avoid burns.

Checklist for skill P PLANNING YOUR WORK	
Candidates:	**Marks awarded**
● plan a simple method to collect evidence	2
● plan to collect evidence that will answer your questions ● plan to use suitable equipment or other ways to get evidence	4
● use scientific knowledge to: – plan and present your method – identify key factors to vary or control – make a prediction if possible ● decide on a suitable number and range of readings (or observations) to collect	6
● use detailed scientific knowledge to: – plan and present your strategy (the approach you have decided on) – aim for precise and reliable evidence – justify your prediction if you made one ● use information from other sources, or from preliminary work in your plan	8

O Obtaining your evidence

Making accurate measurements and observations

Accuracy is important, as well as taking care in checking your results.

Common mistakes include:
- not checking your balance is on zero when measuring mass,
- forgetting to record starting temperatures,
- using a Newton meter with the wrong range to measure forces,
- connecting ammeters into circuits the wrong way around.

Checklist for skill O OBTAINING YOUR EVIDENCE	
Candidates:	**Marks awarded**
• use simple equipment safely to collect some results	2
• make adequate observations or measurements to answer your questions • record the results	4
• make observations or measurements, – with sufficient readings, – which are accurate, and – repeat or check them if necessary • record the results clearly and accurately	6
• carry out your practical work – with precision and skill, – to obtain and record reliable evidence, – with a good number and range of readings	8

You should consider if using **data-loggers** will improve the quality of the evidence you collect.

If one of your results seems unusual, make sure you repeat it. If you find that it was an error, you don't have to include it in your final results.
(But do comment on it in your evaluation!)

Recording your results

You will often record your results in a table.
In the first column of your table you put the thing (variable) that you changed in your experiment.
In the second column you put the thing (variable) that you judged or measured.
Look at the examples below:

Electromagnet	Number of paper clips picked up
30 turns of wire	5
45 turns of wire	14
60 turns of wire	32
75 turns of wire	58

You changed the number of turns on the electromagnet.   *You measured how many paper clips it picked up.*

Voltage across bulb (volts)	Current through bulb (amps
0	0
2	0.5
4	2
6	1.5
8	1.75

You changed the voltage across the bulb.   *You measured the current through the bulb.*

A   Analysing and drawing conclusions

Drawing graphs and bar charts

Once you have recorded your results in a table,
drawing a graph will show you any patterns.

Whether you draw a bar chart or a line graph
depends on your investigation.
Here's a quick way to decide which to draw:

> If the thing (variable) you change is described in words,
> then draw a **bar chart**.
> If the thing (variable) you change is measured,
> then draw a **line graph**.

Let's look at the results from the tables on the previous page:

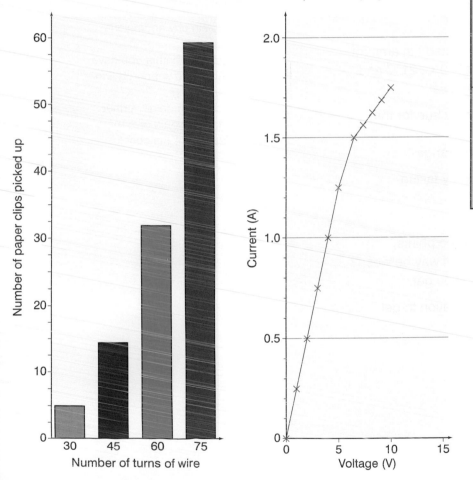

Number of turns of wire

Voltage (V)

Notice that the thing (variable) you change always goes along
the bottom of your graph. The thing (variable) you measure
goes up the side.

You can then see if there is a link between the two variables.
For example, the greater the number of turns of wire there were
on the electromagnet, the more paper clips it picked up.
See the line graph above.

Then try to explain any patterns you spot on your graphs
using the ideas you have learned about in science.

E   Evaluating

When you have drawn your conclusions, you should now think about how well you did your investigation.

Ask yourself these questions to see if you could have improved your investigation:

- Were my results accurate?

- Did any seem 'strange' compared to the others?
  These are called **anomalous** results.

- Should I have repeated some tests to get more reliable results?
  Could I improve the method I used?

- Did I get a suitable range of results?
  The range is the spread of values you chose.

  For example:
  if you were seeing how temperature affected the amount a ball bounced, choosing to do tests at 20°C, 21°C and 22°C would not be a good range to choose!

- If there is a pattern in my results, is it only true for the range of values that I chose?
  Would the pattern continue beyond this range?

- Would it be useful to check your graph by taking readings *between* points?

  For example:
  if you have a sudden change between two points,
  why not do another test to get a point half way between?
  This would check the shape of the line you get.

- How would I have to change my investigation to get the answers to the questions above?

Checklist for skill E EVALUATING YOUR EVIDENCE	
Candidates:	**Marks awarded**
• make a relevant comment about the method used or the results obtained	2
• comment on the accuracy of the results, pointing out any anomalous ones • comment on whether the method was a good one and suggest changes to improve it	4

• look at the evidence and:
  – comment on its reliability,
  – explain any anomalous results,
  – explain whether you have enough to support a firm conclusion
• describe, in detail, further work that would give more evidence for the conclusion
	6

Drill the skill and crack Sc1.

As you study Science, you will need to use some general skills along the way.
These general learning skills are very important, whatever subjects you take or job you go on to do.

The Government has recognised just how important the skills are by introducing a new qualification.
It is called the **Key Skills Qualification**.
There are 6 key skills:

- **Communication**
- **Application of number**
- **Information Technology (IT)**
- **Working with others**
- **Problem solving**
- **Improving your own learning**

The first 3 of these key skills will be assessed by exams and by evidence put together in a portfolio.
You can see what you have to do to get the first level in the sections below.

Communication

In this key skill you will be expected to:

- **hold discussions**
- **give presentations**
- **read and summarise information**
- **write documents**

You will do these as you go through your course and producing your coursework will help.

Look at the criteria below:

> *What you must do ...*
> Take part in discussions.
> Read and obtain information.
> Write different types of document.

Application of number

In this key skill you will be expected to:

- **Obtain and interpret information**
- **Carry out calculations**
- **Interpret and present the results of calculations**

> *What you must do ...*
> Interpret information from different sources.
> Carry out calculations.
> Interpret results and present your findings.

Information Technology

In this key skill you will be expected to:

- **Use the internet and CD ROMs to collect information**
- **Use IT to produce documents to best effect**

> *What you must do ...*
> Find, explore and develop information.
> Present information, including text, numbers and images.

▶▶▶ Revising and doing your exams

When you walk into your Science exam, you will already have your coursework marks completed.
If you do Modular Science, you will also have your test marks.
But your final exam is still the biggest part of your GCSE.
So it's important that you prepare well and feel good on the day.

Plan your revision in the weeks leading up to the exams.
Don't leave it too late!

The question 1's at the end of each chapter are a good way to revise the Summaries. These contain the essential notes you need.
Then try the past paper questions (coloured pages) on the chapter.
If you get stuck, ask a friend or your teacher the next day for help.

Just sitting there (especially in front of the TV!), reading your notes isn't good enough for most people. *Active* revision is better.
And don't try to revise for too long without a break.
Do 25 minutes, then promise yourself a 10 minute rest.
This works better than trying to revise non-stop.

So you've finished your revision (it's too late to worry about that anyway!), and it's the day of the exam. What will you need?
Remember to bring:

- Two pens (in case one runs out).
- A pencil for drawing diagrams.
- An eraser and ruler.
- A watch for pacing yourself during the exam.
  (It might be tricky to see the clock in the exam room.)
- A calculator (with good batteries).

You will feel better if you know exactly what to expect.
So collect all the information about your exam papers.
You can use a table like the one shown below:

Date, time and room	Subject, paper number and tier	Length (hours)	Types of question: – structured? – single word answers? – longer answers? – essays?	Sections?	Details of choice (if any)	Approximate time per page (minutes)
4th June 9.30 hall	Science (Double Award) Paper 2 (Physics) Foundation Tier	1½	Structured questions (with single-word answers and longer answers)	1	no choice	4–6 min.

▶▶▶ Equations glossary

Equation 1 (page 46)

work done = energy transferred

Equation 2 (page 46)

work done = force applied × distance moved in direction of force

Joule (J)   Newton (N)   Metre (m)

Equation 3 (page 48)

change in gravitational = weight × height potential energy

Joule (J)   Newton (N)   Metre (m)

Equation 4 (page 50)

$$\text{power} = \frac{\text{work done (joules, J)}}{\text{time taken (seconds, s)}}$$

(watt, W)

This equation is often written to calculate energy transfer instead:

energy transferred = power × time

Equation 5 (page 52)

Electrical energy

energy transferred = power × time

kilowatt hour (kWh)   kilowatt (kW) × hours (h)

Equation 6 (page 53)

Cost of electrical units

total cost = number of units × cost per unit

(£ and pence)   (kWh used)   (pence each)

Equation 7 (page 54)

$$\text{efficiency} = \frac{\text{useful energy transferred by device}}{\text{total energy supplied to device}}$$

Equation 8 (page 57)

kinetic energy = $\frac{1}{2}$ × mass × (velocity)^2 (squared)

Symbol form: KE = $\frac{1}{2}$ mv^2

Equation 9 (page 67)

Power in electrical circuits

power = voltage × current

P   =   V   ×   I

Watts (W)   Volts (V)   Amps (A)

Equation 10 (page 68)

Ohm's law

voltage (pd) = current × resistance

V   =   I   ×   R

volts (V)   ampere (A)   ohm (Ω)

Equation 11 (page 125)

Weight = mass × g (gravitational field strength)

newton, (N)   kg   Newton per kilogram, N/kg

Equation 12 (page 130)

$$\text{Average speed} = \frac{\text{distance travelled (metres, m)}}{\text{time taken (seconds, s)}}$$

(m/s)

Equation 13 (page 139)

$$\text{acceleration} = \frac{\text{change in velocity (m/s)}}{\text{time taken for change seconds (s)}}$$

metres per second squared (m/s^2)

This equation is often used in this symbol form:

$$a = \frac{(v - u)}{t}$$

where a = acceleration
v = final velocity
u = original velocity
t = time taken for change

Equation 14 (page 171)

wave speed = frequency × wavelength (λ)

(m/s)   Hertz (Hz)   metres (m)

Moments (page 126)

Turning force = Force × perpendicular distance to pivot

(Nm)   Newton (N)   metres (m)

Balanced turning forces (page 126)
(Balanced moments)

Sum of clockwise = sum of anti-clockwise
moments   moments

Pg	Question	Answer
9	3	400
47	b)	200 words
47	c)	25 J
	d)	15 J
47	e)	100 000 words
47	2	200 000 J
47	3 (i)	500 N
49	e)	10 J
49	f)	90 J
49	2	12 m
49	3	1500 J
51	c)	100 W
51	d)	200 W
51	2i)	1500 J
51	2ii)	90 000 J
51	2iii)	540 000 J
52	b)	3 600 000 J
53	d)	32p
53	e)	The same
53	2 Light	1.5 kWh, 12p
53	2 Oven	6 kWh, 48p
53	2 Mower	0.5 kW, 1 kWh
53	2 TV	3 kWh, 24p
53	2 Stereo	10 h, 2 kWh
55	2 TV	20% or 0.2
55	2 Toaster	400 J
55	2 Computer	50% or 0.5
55	2 Light bulb	5 W
57	di)	900 J
57	dii)	3600 J
57	diii)	8100 J
57	3 (1 m/s)	1 J
57	3 (2 m/s)	4 J
57	3 (4 m/s)	16 J
57	3 (8m/s)	64 J
57	3 (16m/s)	256 J
59	1i)	250 J
59	1ii)	2 m
59	1iii)	80 N
59	2i)	240 J
59	2ii)	700 000 J
59	2iii)	100 N
59	3 Josh	750 J, 15 W
59	3 Chris	600 J, 10 W
59	3 Leanne	400 J, 10 W
59	3 Steven	2800 J, 20 W
59	3 Tammy	48 J, 24 W
59	4i)	0.25 kWh
59	4ii)	1.5 kWh
59	4iii)	2 kW
59	5i)	16p, 4p
59	5ii)	2 h, 8 h
59	5iii)	4i) 2p, 0.5p
59	5iii)	4ii) 12p, 3p
59	5iii)	4iii) 8p, 2p
59	6i)	0.25 or 25%
59	6ii)	65 J
59	7	Both the same
67	2 Kettle	1380 W
67	2 Headlamp	5 A
67	2 Light bulb	6 V
69	2i)	12 V
69	2ii)	0.5 A
73	c)	6 V
73	2i)	0.33 A
73	2ii)	3: 1 V, 6: 2 V, 9: 3 V
79	2i)	460 W
79	2ii)	6.5 A
79	2iii)	60 W
79	3i)	16 V
79	3ii)	0.02 A
79	3iii)	12 ohms
109	d) TV	2 A current, 3 A fuse
109	d) Computer	3 A current, 5 A fuse
109	d) Vacuum	4 A current, 5 A fuse
109	d) Fire	12 A current, 13 A fuse
115	4i)	3 A fuse
115	4ii)	5 A fuse
115	4iii)	13 A fuse
115	4iv)	3 A fuse
125	e)	6.35 kg
125	2	Horse = 2540 N
126	a)	10 Nm
126	b)	2 Nm
126	c)	150 Nm
126	e)	1st row, 30 Ncm
126	e)	2nd row, 24 Ncm, 6 cm
126	e)	3rd row, 3 cm, 6 cm, 36 Ncm
127	g)	2 cm
129	4i)	20 N
129	4ii)	3.2 N
129	4iii)	28 N
129	4iv)	548 N
129	5i)	60 J
129	5ii)	80 J, 12.8 J, 112 J, 2192 J
129	6i)	30 Nm
129	6ii)	3 m
129	6iii)	A = 2 N
129	7i)	8 cm
129	7ii)	14 cm
129	7iii)	30 cm
129	7iv)	1 cm
131	c)	4 m/s
131	d)	10 s
132	c)	4 m
132	d)	1 m
133	2i)	4 m at 4 sec, 6 m at 8 sec
133	2ii)	10 to 13 seconds
133	2iii)	0 m
137	e)	8 m/s
139	e)	5 m/s^2
141	2) Y10 boy	1 m/s
141	2) Milk float	200 m
141	2) Racing flea	2 s
141	2) Helicopter	250 s
141	2) Space liner	300 000 m
141	5i)	Both 18 s away from shop
141	5ii)	They would meet at shop
141	6ii)	10 m/s
143	e)	20 J
148	d)	Car, 364 500 J, Bus 640 000 J
149	f)	12 800 N
149	g)	60.75 m
152	a)	365¼ days
152	b)	24 hours, one day
153	d)	27.4 times greater
159	c)	2000 years ago
170	a)	5 Hz
170	b)	1200 waves
170	c)	2 cm, 1 cm
171	e)	12 m/s
171	f)	2 cm and 4 cm
171	3i)	2 m/s
171	3ii)	680 Hz
172	a)	30°
172	b)	45°
172	c)	0°
177	4i)	2 m/s
177	4ii)	0.5 m/s
177	4iii)	1 m/s
177	4iv)	0.25 Hz
177	5i)	3 m/s
177	5ii)	0.5 Hz
177	ai)	680 Hz
177	aii)	340 Hz
177	aiii)	170 Hz
196	c)	87%
203	d)	10
203	e)	2
203	2ii)	6
203	2iii)	U-235 = 143, U-238 = 146
203	2iv)	10 g or 0.01 kg
204	a)	500 carbon-14 atoms
204	b)	2 years
206	4i)	2000 carbon-14 atoms
206	4ii)	500 carbon-14 atoms

▶▶▶ Acknowledgements

I would like to thank the following people for their help and support in writing this book:

Roger Alkroyd, Nick Blaker, Nick Cornell, Sarah Coulson, Michael Cotter, Michael Goodwin, Beth Hutchins, Stewart Miller, Nick Paul, Frances Pinsent, Henry Pinsent, Pat Pinsent, Stephanie Pinsent, Lawrie Ryan, Sarah Ryan, Steve Sims, Harry Venning and Susannah Wills.

Also the following students from Sheringham High School:
Courtenay-Ann Bobyk, Liam Cooper, Sophie Shaw-Foucher, Jennifer Prince and Kathryn Pumphrey.

Acknowledgement is made to the following Awarding Bodies for their permission to reprint questions from their examination papers:

AQA — Assessment and Qualifications Alliance
AQA (NEAB) — Assessment and Qualifications Alliance
EDEXCEL — Edexcel Foundation
OCR — Oxford, Cambridge and RSA Examinations
WJEC — Welsh Joint Education Committee

Illustration acknowledgements

Allsport: 144t; **Art Directors and Trip:** 34, 56b, 68 both, 103, 111, 114t, 146, 182t, 183ca, 189b, 192c; **Richard Austin,** Lyme Regis, Dorset: 168; **John Cleare Mountain Camera Picture Library:** 42; **Collections:** 14, 15b, 49, 52, 82; **Construction Photography:** 74; **Corbis UK Ltd:** 7, 63, 101, 183b; **Education Photos/John Walmsley:** 57, 113; **Empics:** 144b; **Eye Ubiquitous:** 13, 22t, 26, 67, 107, 114b, 154, 189t, 190; **Flight International Collection at Quadrant Picture Library:** 191t; **Getty Images/Image Bank:** 145b, 170; **Getty Images/Stone:** 167; **Getty Images/Telegraph Colour Library:** 198; **Robert Harding Picture Library:** 22b, 145t; **Impact Photos:** 187t; **Jeff Moore (jeff@jmal.co.uk):** 24, 41, 76, 84, 85, 88, 94, 109, 143, 147, 174, 179, 183t, 193; **NASA:** 119; **PA News Photos:** 29, 169; **Popperfoto:** 56t, 138; **Rex Features:** 12, 15t, 86, 123; **Science Photo Library:** 11, 21, 27, 37 both, 47, 86, 134 both, 152, 155, 156, 157 all, 160, 180, 182b, 183cb, 187b, 188, 191b, 192t and b, 199, 200, 201, 204.

While every effort has been made to contact copyright holders, the publishers apologise for any omissions, which they will be pleased to rectify at the earliest opportunity.

Picture research by Liz Moore (lm@appleonline.net)

Grateful thanks to Hannah Sherry, John Gibson, Nigel Dowrick and the Physics Department at Benenden School for their help in producing some photographs and to Sean Davies for enduring the rigours of the Shock Machine at Hastings Flamingo Arcade (page 88).